MW00844118

Structural Loads Analysis for Commercial Transport Aircraft: Theory and Practice

Structural Loads Analysis for Commercial Transport Aircraft: Theory and Practice

Ted L. Lomax

EDUCATION SERIES

J. S. Przemieniecki
Series Editor-in-Chief
Air Force Institute of Technology
Wright-Patterson Air Force Base, Ohio

Published by
American Institute of Aeronautics and Astronautics, Inc.
1801 Alexander Bell Drive, Reston, VA 22091

American Institute of Aeronautics and Astronautics, Inc., Reston, Virginia

Library of Congress Cataloging-in-Publication Data

Lomax, Ted L.
    Structural loads analysis for commercial transport aircraft :
  theory and practice / Ted L. Lomax.
      p. cm. — (AIAA education series)
  Includes bibliographical references and index.
    1. Airframes—Design and construction.   2. Structural dynamics.
  3. Transport planes—Design and construction.   I. Title.
  II. Series.
  TL671.6.L597   1995   629.134$'$31—dc20   95-22159
  ISBN 1-56347-114-0

Second Printing

Copyright © 1996 by the American Institute of Aeronautics and Astronautics, Inc. All rights reserved. Printed in the United States. No part of this publication may be reproduced, distributed, or transmitted, in any form or by any means, or stored in a database or retrieval system, without the prior written permission of the publisher.

Data and information appearing in this book are for informational purposes only. AIAA is not responsible for any injury or damage resulting from use or reliance, nor does AIAA warrant that use or reliance will be free from privately owned rights.

Texts Published in the AIAA Education Series

Texts Published in the AIAA Education Series (continued)

Space Vehicle Design
    Michael D. Griffin and James R. French, 1991
Introduction to Mathematical Methods in Defense Analyses
    J. S. Przemieniecki, 1990
Basic Helicopter Aerodynamics
    J. Seddon, 1990
Aircraft Propulsion Systems Technology and Design
    Gordon C. Oates, Editor, 1989
Boundary Layers
    A. D. Young, 1989
Aircraft Design: A Conceptual Approach
    Daniel P. Raymer, 1989
Gust Loads on Aircraft: Concepts and Applications
    Frederic M. Hoblit, 1988
Aircraft Landing Gear Design: Principles and Practices
    Norman S. Currey, 1988
Mechanical Reliability: Theory, Models and Applications
    B. S. Dhillon, 1988
Re-Entry Aerodynamics
    Wilbur L. Hankey, 1988
Aerothermodynamics of Gas Turbine and Rocket Propulsion,
    Revised and Enlarged
    Gordon C. Oates, 1988
Advanced Classical Thermodynamics
    George Emanuel, 1988
Radar Electronic Warfare
    August Golden Jr., 1988
An Introduction to the Mathematics and Methods of Astrodynamics
    Richard H. Battin, 1987
Aircraft Engine Design
    Jack D. Mattingly, William H. Heiser, and Daniel H. Daley, 1987
Gasdynamics: Theory and Applications
    George Emanuel, 1986
Composite Materials for Aircraft Structures
    Brian C. Hoskins and Alan A. Baker, Editors, 1986
Intake Aerodynamics
    J. Seddon and E. L. Goldsmith, 1985
Fundamentals of Aircraft Combat Survivability Analysis and Design
    Robert E. Ball, 1985
Aerothermodynamics of Aircraft Engine Components
    Gordon C. Oates, Editor, 1985
Aerothermodynamics of Gas Turbine and Rocket Propulsion
    Gordon C. Oates, 1984
Re-Entry Vehicle Dynamics
    Frank J. Regan, 1984

Published by
American Institute of Aeronautics and Astronautics, Inc., Reston, Virginia

Foreword

As one of its major objectives, the AIAA Education Series is creating a comprehensive library of the established practices in aerospace design. *Structural Loads Analysis for Commercial Transport Aircraft: Theory and Practice*, by Ted L. Lomax, provides an authoritative exposition of load analysis theories and practice as applied to structural design and certification. In writing this text, the author has captured years of experience in the field as a structural loads engineer and manager at the Boeing Company on several different types of commercial transport aircraft.

The 16 chapters in this text are arranged into topics dealing with maneuvering and steady flight loads (symmetrical flight, rolling, yawing, turbulence), landing and gust loads, aircraft component loads (horizontal and vertical tail, wing, body, control surfaces, and high-lift devices), aeroelastic considerations (flutter, divergence, and control reversal), structural design considerations, and design airspeeds. Each chapter provides some simplified approaches to verify computer-generated analyses, thereby providing additional confidence that the work is correct. These approaches also add to a better understanding of the various parameters influencing modern designs.

The AIAA Education Series embraces a broad spectrum of theory and application of different disciplines in aeronautics and astronautics, including aerospace design practice. The series also includes texts on defense science, engineering, and technology. It provides both teaching texts for students and reference materials for practicing engineers and scientists. *Structural Loads Analysis for Commercial Transport Aircraft: Theory and Practice* will be a valuable resource for aircraft design teams. It complements several other texts on aircraft design previously published in the series.

J. S. Przemieniecki
Editor-in-Chief
AIAA Education Series

Table
of
Contents

Preface

Structural loads analyses have come a long way since the early days of commercial aviation when the work was done with slide rules and desk calculators.

We have moved into the age of computers that can do wonders by managing large amounts of data, creating enormous databases, and giving minute details in various structural components; we can even go from concept to hardware without a drawing.

The question is, **"Is it right?"**

One of the purposes of this book is to provide some simplified approaches whereby checks may be applied to more elaborate analyses, thereby providing some confidence that the work is correct. The use of simplified analysis techniques will allow engineers to better judge the correctness of their work, thus producing a well-designed product, which in the end is the purpose of all our work.

The other purpose of this book is to provide a compendium of various loads analyses theories and practices as applied to the structural design and certification of commercial transports certified under the Federal Aviation Regulations Part 25.

In general, these discussions will be related to the work the author has accomplished and experienced during his fortysome years as a structural loads engineer and manager at The Boeing Company.

It is not the intention of the author that these discussions be used as a reference for current application of the regulations applied to a given model but rather that they provide only a historical record of how loads analysis theory and practice have changed over the years from 1953 to the present.

I hope that this book will provide some continuity between what was done on earlier aircraft designs and what the current applications of the present regulations require, and hence that it will be of use to younger load engineers in understanding and applying good engineering practice to new designs in the future.

Acknowledgment

I am thankful to the Boeing Company Structures Department management, particularly J. A. McGrew and R. M. Thomas, for their encouragement and support. I appreciate the contributions of Ed Lamb and Bob Martin for their assistance and recommendations on technical content. The committee of young engineers who reviewed and critiqued the early development stages of the book gave me insight on format and technical depth.

I am grateful and appreciative to my wife, Gloria, and my family who encouraged and supported me, and proofread the text.

Ted L. Lomax
1996

Nomenclature

The nomenclature shown in this section are general in nature. Specific symbols are explained as required in each chapter.

BS	= body coordinate station
BTL	= balancing tail load at $M_t = 0$, lb
b	= wing reference span, ft
C_D	= drag coefficient
C_F	= chord force, lb
CG	= airplane center of gravity position, (% mac/100)
C_L	= tail-off lift coefficient, $L/(qs_w)$
C_{La}	= total airplane lift coefficient
C_{Lmax}	= maximum lift coefficient
C_M	= pitching moment coefficient
$C_{M0.25}$	= pitching moment coefficient about 0.25 mac wing, $M_{0.25}/(q S_w c_w)$
C_N	= normal force coefficient
C_{Nmax}	= maximum normal force coefficient
C_n	= yawing moment coefficient
C_l	= rolling moment coefficient
C_y	= side force coefficient
c_w	= wing mean aerodynamic chord, in.
G	= gust gradient, chords or ft
g	= acceleration of gravity, ft/s^2
I_x, I_y, I_z	= moment of inertia in roll, yaw, and pitch, slug ft^2
L	= aerodynamic lift, lb
L_t	= horizontal tail load, lb
M	= Mach number
$M_{0.25}$	= aerodynamic pitching moment about 0.25 mac wing, in.-lb
M_t	= horizontal tail pitching moment, in.-lb
mac	= mean aerodynamic chord, in.
N_F	= normal force, lb
n_x	= longitudinal load factor in the x axis
n_y	= lateral load factor in the y axis
n_z	= vertical load factor in the z axis
q	= dynamic pressure, lb/ft^2
S_w	= wing reference area, ft^2
T_{eng}	= engine thrust, lb
V_A	= design maneuver airspeed, knots equivalent airspeed (**keas**)
V_B	= design gust airspeed for maximum gust velocity, keas
V_C	= design cruise speed, keas
V_c	= calibrated airspeed, knots calibrated airspeed (kcas)

V_D = design dive speed, keas
V_{+HAA} = airspeed at the upper left-hand corner of the V-n diagram
V_i = indicated airspeed, knots indicated airspeed (kias)
V_t = true airspeed, knots true airspeed (ktas) or ft/s
W = airplane gross weight, lb
x, y, z = airplane reference axes, see Fig. 1.1
x_t = distance between 0.25 mac wing and 0.25 mac horizontal tail, in.
z_e = engine thrust coordinate, in.
α_w = wing angle of attack, deg
β = airplane sideslip angle, deg
δ = control surface deflection and pressure ratio of the atmosphere
θ = airplane pitch angle, deg
ρ = density of air, slug/ft^3
σ = density ratio, ρ/ρ_0
ϕ = airplane roll angle, deg
ψ = airplane yaw angle, deg

Time Derivative Convention
$\dot{\phi}$ = $d\phi/dt$, rad/s
$\ddot{\phi}$ = $d^2\phi/dt^2$, rad/s^2

Subscripts
A = total airplane
o = sea level condition
r = rudder
ss = steady sideslip
w = wheel angle
α = angle of attack
β = sideslip

1
Introduction

Structural load analysis implies the calculation or determination of the loads acting on the aircraft structure for flight maneuvers, flight in turbulence, landing, and ground-handling conditions.

The present methods used to determine those loads may be complex and involve the use of advance technologies in which the total airplane is solved as a complete system using large digital computers. Loads are applied to all of the major structural components of the aircraft in the form of panel aerodynamic and inertia loads that require solution of multidegrees of freedom when considering the effects of structural dynamics on the airplane response.

Because of the magnitude of the number of load points and conditions that may be investigated, the necessity of validating the results becomes an important and time-consuming task. In the "olden days" when structural analyses were less complex, the ability to determine structural loads was relatively simple, even though computers were used.

1.1 Applicability of the Analysis

The structural load analyses discussed in this book are applicable for the determination of 1) design load conditions, 2) fail-safe load conditions, 3) fatigue load analyses; and 4) operating load conditions.

1.2 Criteria

The criteria discussed are for commercial aircraft designed up to the time this book was written. Those criteria were taken from the United States and European regulations and from the joint European/United States harmonization working group.[1-3]

Even though the discussions and methods of analysis are based on the criteria discussed in this section, the methodology may be applied to aircraft designed to other criteria.

In general, only passing reference is made to military aircraft criteria or analysis methods except in the adoption of a specific method by civilian authorities where previous methods were not acceptable.

1.3 Methodology

As stated in the opening paragraphs of this chapter, aircraft load analyses have become complex in nature and require very sophisticated computing systems to solve the resulting equations of motion for the aircraft.

Two of the main purposes of this book are 1) to provide a historical background of the philosophy of the criteria, methods, and practice used for structural loads analyses since the conception of the DC-8 and 707 aircraft in the early 1950s and 2) to provide simplified analytical methods and approaches for calculating structural analysis loads that will allow engineers to make quick checks of the loads obtained from more sophisticated analyses; to determine the criticality of one condition vs another such as gross weight, fuel distribution and usage, airspeed, or Mach number effects; and to assess growth potential for an aircraft by varying airplane center of gravity, gross weights, and airspeeds.

1.3.1 Static Load Analyses

The static load analyses methods and equations discussed in this book reflect the experience of the author and therefore should not be assumed as the only way to solve for a particular set of loads.

Each aircraft may have a particular configuration that requires the inclusion of significant parameters that have been neglected in the equations shown in this book. An example would be the inclusion of thrust effects for an aircraft with body-mounted engines with a high thrust line, thus increasing the downtail load for forward center of gravity positions. Those effects, although provided in the derivation of the analysis, have been neglected in the simplified equations.

The equations and methods of analyses shown in this book need to be modified to fit the configurations under investigation.

1.3.2 Dynamic Load Analyses

Although dynamic load analysis results are shown in various parts of the book, the methods of analysis for determining dynamic loads due to flight in turbulence or while landing or taxiing are not discussed in detail. The inclusion of significant structural degrees of freedom along with the representation of flight control augmentation systems requires significant mathematical modeling to adequately represent the airplane.

The references at the end of this chapter, shown for historical purposes, provide sources that have been used in developing dynamic load analyses methods.[4-7]

1.4 Static Aeroelastic Phenomena

The regulations specifically require that if deflections under load would significantly change the distribution of external and internal loads, the redistribution must be taken into account, per FAR 25.301(c). Therefore, the static aeroelastic phenomena discussed in this book are 1) the effect of static aeroelasticity on resulting structural loads for the wing and empennage; 2) the inclusion of aeroelastic effects on the stability derivatives required for solution of the equations of motion for maneuvers in pitch, roll, and yaw and flight in turbulence; and 3) the evaluation of the static divergence and reversal characteristics of the wing and empennage due to aeroelasticity.

1.5 Sign Convention

The sign convention is shown in Fig. 1.1 for the analyses presented in this book.

Mass data such as airplane gross weight and moments of inertia are represented with respect to the aircraft center of gravity for the specific condition under investigation.

Aerodynamic pitching, rolling, and yawing moments are represented with respect to the quarter-chord reference of the airplane wing, unless stated differently, such as the horizontal tail parameters.

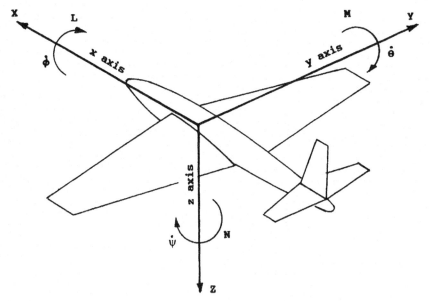

Fig. 1.1 Sign convention. L = rolling moment; M = pitching moment; N = yawing moment; X, Y, Z = components of resultant aerodynamic forces; $\dot{\phi}$ = roll rate; $\dot{\Theta}$ = pitch rate; $\dot{\psi}$ = yaw rate.

References

[1]Anon., "Part 25—Airworthiness Standards: Transport Category Airplanes," Federal Aviation Regulations, U.S. Dept. of Transportation, Jan. 1994.

[2]Anon., "JAR 25 Large Aeroplanes," Joint Aviation Requirements, Oct. 1989.

[3]Anon., "Loads Harmonisation Working Group Recommendations," Joint Aviation Proposal, March, 1993.

[4]Bisplinghoff, R. L., Ashley, H., and Halfman, R. L., *Aeroelasticity,* Addison–Wesley, Reading, MA, 1955.

[5]Fung, Y. C., *An Introduction to the Theory of Aeroelasticity,* Wiley, New York, 1955.

[6]Hoblit, F. M., *Gust Loads on Aircraft: Concepts and Applications,* AIAA Education Series, AIAA, Washington, DC, 1989.

[7]Miller, R. D., Kroll, R. I., and Clemmons, R. E., "Dynamic Loads Analysis System (Dyloflex) Summary," NASA CR-2846, 1978.

Symmetrical Maneuvering Flight

2.1 Symmetrical Maneuvering Flight Definition

Symmetrical flight conditions are defined in this book as flight maneuvers about the lateral (pitch) axis of the airplane in which only lift and pitch are considered. The assumptions made for analytical purposes are that 1) airspeed and Mach number (hence altitude) are constant during the maneuver and that 2) aircraft roll and yaw perturbations are neglected or assumed zero during the maneuver.

2.1.1 Symmetrical Flight Conditions

Symmetrical flight conditions would normally include any maneuver for which the aircraft is to be designed that does not involve motion about the roll or yaw axis. Since the subject of this book pertains to structural load analysis for commercial transport aircraft, symmetrical flight loads will be considered only for the following maneuvers or conditions: 1) steady-state flight conditions such as those shown in Fig. 2.1 for wind-up turns and roller coaster maneuvers and 2) abrupt pitching maneuvers as shown in Fig. 2.2 for the unchecked up elevator condition and the elevator checkback condition at design load factors.

2.1.2 Parameters Required for Load Analysis

The solution to the symmetrical flight maneuver analyses discussed in this chapter will provide the following data that are required for determination of body, horizontal tail, nacelle, and wing loads: 1) wing reference angle of attack α_w, 2) horizontal tail loads Lt and Mt and elevator angle δ_e required for the maneuver, 3) rate parameters $\dot{\alpha}$ and $\dot{\theta}$, and 4) pitching acceleration $\ddot{\theta}$.

2.2 Symmetrical Maneuver Load Factors

Except where limited by maximum static lift coefficients, the airplane is assumed to be subjected to symmetrical maneuvers resulting in the limit maneuvering load factors per FAR 25.337(b) and (c) and FAR 25.345(a)(1) and (d), shown in Tables 2.1 and 2.2.

Table 2.1   Limit design load factors for flaps up

Airspeeds	Up to V_C	At V_D
Positive maneuvers[a]	2.5	2.5
Negative maneuvers	−1.0	0

[a] see Eq. (2.1).

Table 2.2 Limit design load factors for flaps down

Flap position	Gross weight	Load factors
Takeoff	Maximum takeoff	2.0 and 0
Landing	Maximum landing	2.0 and 0
Landing	Maximum takeoff	1.5 and 0

For gross weights less than 50,000 lb,

$$n_z = [2.1 + 24,000/(W + 10,000)] < 3.8 \, \text{max} \qquad (2.1)$$

Symmetrical maneuvering load factors lower than those shown in Table 2.1 may be used if the airplane has design features that make it impossible to exceed these values in flight, per FAR 25.337(d). An example of such a design feature

a)

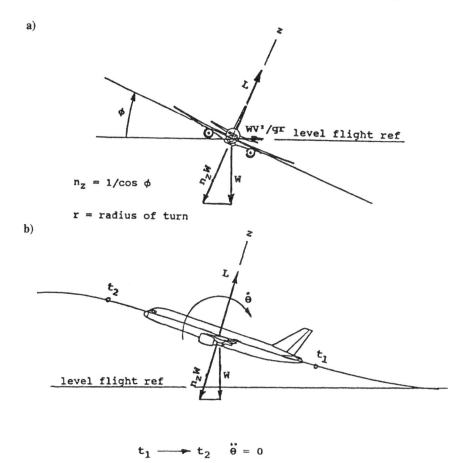

b)

Fig. 2.1 Steady-state maneuvers: a) wind-up turn and b) roller-coaster maneuver.

would be a "black box" with redundant fail-safe backup that limits the maneuver load factor to a given selected fixed value.

2.3 Steady-State Symmetrical Maneuvers

Steady-state symmetrical maneuvers are defined as conditions in which the pitching acceleration is assumed negligible or zero.

The wind-up turn as shown in Fig. 2.1 is considered a steady-state symmetrical condition, even though the airplane does have an acceleration acting laterally during the turn. The roller coaster maneuver, if accomplished slowly with respect to the change in pitch rate, may be flown with negligible or zero pitching acceleration.

2.3.1 Steady-State Symmetrical Maneuver Equations

The normal and chord forces acting on the airplane, as shown in Fig. 2.3, may be determined from the summation of forces in the z and x axes:

$$N_F = n_z W \tag{2.2}$$

$$C_F = n_x W + T_{\text{eng}} \tag{2.3}$$

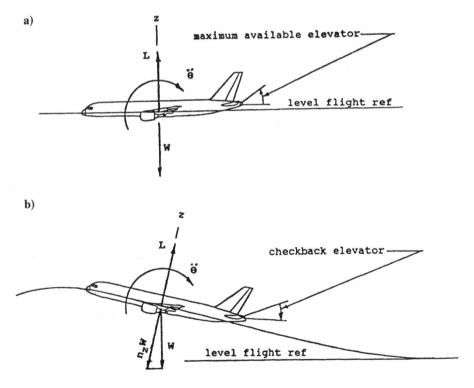

Fig. 2.2 Abrupt pitching maneuvers: a) abrupt unchecked elevator condition and b) elevator checkback condition.

The relationship between normal and chord forces and lift and drag, shown in Fig. 2.3, may be determined by Eqs. (2.4) and (2.5):

$$N_F = L \cos \alpha_w + D \sin \alpha_w \qquad (2.4)$$

$$C_F = D \cos \alpha_w - L \sin \alpha_w \qquad (2.5)$$

Using the simplification that the normal force is equal to lift, and that the lift and pitching moments may be considered as the sum of the tail-off plus the horizontal tail loads as shown in Fig. 2.4, then one can derive the lift and pitch balance equations with respect to the 0.25 of the mean aerodynamic chord:

$$L + L_t = n_z W \qquad (2.6)$$

$$M_{0.25} + n_z W x_a + n_x W z_a = L_t x_t - M_t - T_{eng} z_e \qquad (2.7)$$

The horizontal tail drag term is neglected in Eq. (2.7) as small with respect to the effect on airplane pitching moment. This assumption may not be valid for aircraft configurations with horizontal tails mounted on the vertical tail, such as the BAC 111 and 727 aircraft. The effect of neglecting the horizontal tail drag in Eq. (2.7) is shown in Table 2.3.

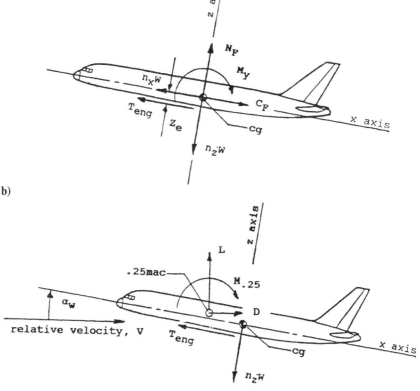

Fig. 2.3 Forces acting on the airplane during steady-state symmetrical maneuvers (pitching acceleration = 0).

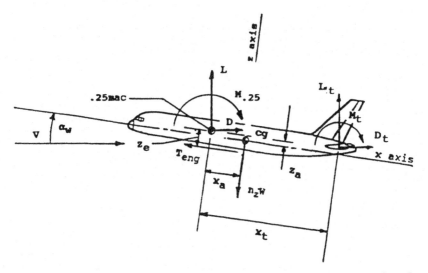

Fig. 2.4 Balancing tail loads during steady-state symmetrical maneuvers: where $L =$ airplane tail-off lift, lb; $D =$ airplane tail-off drag, lb; $M_{0.25} =$ tail-off pitching moment about 0.25 mac, in.lb; $L_T =$ horizontal tail load, lb; $D_T =$ horizontal tail drag, lb; $M_T =$ horizontal tail pitching moment about the tail reference axis, in.lb.

The following relationships are defined:

$$x_a = (CG - 0.25)c_w \tag{2.8}$$

$$C_{La} = n_z W / q S_w \tag{2.9}$$

$$M_{0.25} = C_{M0.25} q S_w c_w \tag{2.10}$$

Inserting Eqs. (2.8) and (2.9) into Eq. (2.7), and combining Eqs. (2.6) and (2.7), one can determine the balancing tail load:

$$L_t = [(CG - 0.25)C_{La} + C_{M0.25}]q S_w c_w / x_t \\ + M_t / x_t + T_{eng} z_e / x_t + n_x W z_A / x_t \tag{2.11}$$

Neglecting the last term in Eq. (2.11) as small, then

$$L_t = [(CG - 0.25)C_{La} + C_{M0.25}]q S_w c_w / x_t \\ + M_t / x_t + T_{eng} z_e / x_t \tag{2.12}$$

A further simplification may be made by neglecting the term M_t / x_t in Eq. (2.12) as small with respect to the pitching moment about the 0.25 mac. Assuming a power-off condition in which thrust is assumed to be zero, one can derive the traditional equation for determining balancing tail load (BTL) in Eq. (2.13):

$$BTL = [(CG - 0.25)C_{La} + C_{M0.25}]q S_w c_w / x_t \tag{2.13}$$

Table 2.3   Effect of neglecting horizontal tail drag in Eq. (2.7) shown for a "T" tail configuration

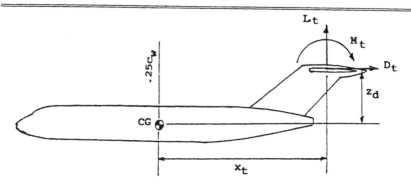

Speed condition	At M_C	V_C at SL	V_F Landing flaps
Altitude, ft	26,600	0	0
V_e, keas	350	350	180
Mach no.	0.90	0.53	Flaps 30
W, lb	169,000	169,000	142,500
CG, $\%c_w/100$	0.10	0.10	0.13
n_z	2.5	−1.0	2.0
Solution neglecting horizontal tail drag			
L_t,[a] lb	−40,100	−3,880	−45,600
M_t, 10^6 in.-lb	2.86	−2.61	1.66
D_t, lb	0	0	0
Solution including horizontal tail drag			
L_t,[a] lb	−38,700	−3,400	−43,800
M_t, 10^6 in.-lb	2.86	−2.61	1.66
D_t lb	3,700	1,300	4,800

[a] $L_t = (L_t)_{DT} + 0.375 D_t$ where $z_d/x_t = 0.375$.

whereby BTL indicates a solution assuming the horizontal tail pitching moment M_t is zero.

The power-on solution may be obtained by including the thrust term in Eq. (2.13):

$$BTL = [(CG - 0.25)C_{La} + C_{M0.25}]q S_w c_w / x_t + T_{eng} z_e / x_t \qquad (2.14)$$

Equating Eqs. (2.12) and (2.14), then

$$BTL = L_T - M_T / x_t \qquad (2.15)$$

The load factor acting along the x axis may be determined from Eq. (2.3):

$$n_x = (C_F - T_{eng})/W \qquad (2.16)$$

$$n_x = (D \cos \alpha_w - L \sin \alpha_w - T_{eng})/W \qquad (2.17)$$

Table 2.4 Substantiation of approximation for balancing tail loads calculated using Eq. (2.15)

Condition	Positive maneuver forward CG	Negative maneuver forward CG	Positive maneuver aft CG
n_z	2.5	−1.0	2.5
W, lb	297,000	297,000	324,000
CG	Forward	Forward	Aft
x_t, in.	1,000	1,000	1,000
Simplified solution			
BTL, lb	−38,000	−12,000	32,000
L, lb	780,500	−285,000	778,000
Exact solution			
L_t, lb	−35,250	−16,120	+31,850
M_t, 10^6 in.-lb	2.75	−4.12	−0.15
L, lb	777,750	−280,880	778,150
Comparison			
$L_{\text{simplified}}/L_{\text{exact}}$	1.004	1.015	1.000

A comparison of the approximate method for calculating balancing tail loads using Eq. (2.15) with the exact solution is shown in Table 2.4. In general, tail loads calculated using the simplified method are conservative for forward center of gravity positions at positive load factors. For aft center of gravity positions, the horizontal tail pitching moments are usually small because the elevator required to produce the maneuver is small; hence the differences between the approximate and exact solutions are small. For negative load factors the solution differences are less than 2% for the example shown.

2.3.2 Solution to Steady-State Maneuver Equations

Inspection of Eqs. (2.13) and (2.14) will show that all of the parameters on the right side of the equations are known, except for $C_{M0.25}$. These equations may be solved using either of the following methods: 1) a solution based on linearized aerodynamic coefficients or 2) a graphical solution when the tail-off pitching moment $C_{M0.25}$ is nonlinear with respect to lift coefficient C_L.

2.3.3 Balancing Tail Loads Using Linearized Coefficients

Assuming linear tail-off aerodynamic coefficients as shown in Fig. 2.5:

$$C_L = C_{Lo} + C_{Law}\alpha_w \qquad (2.18)$$

$$C_{M0.25} = C_{Mo} + \left(\frac{dC_M}{dC_L}\right)C_L \qquad (2.19)$$

a)

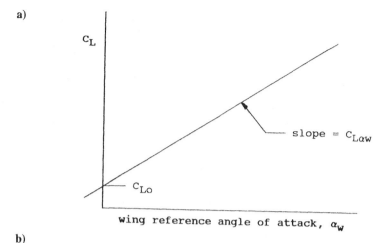

b)

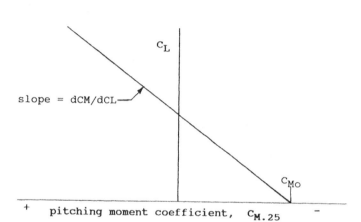

Fig. 2.5   Linear aerodynamic tail-off coefficients: a) see Eq. (2.18); b) see Eq. (2.19).

Inserting Eqs. (2.18) and (2.19) into Eq. (2.13), one can determine the balancing tail load and wing reference angle of attack for the power-off condition:

$$BTL = \frac{[n_z W(CG - 0.25 + dC_M/dC_L) + C_{Mo} q S_w]}{x_t/c_w + dC_M/dC_L} \tag{2.20}$$

$$\alpha_w = \left[\frac{n_z W - BTL}{q S_w} - C_{Lo}\right] \Big/ C_{L\alpha w} \tag{2.21}$$

Differentiating Eqs. (2.20) and (2.21) with respect to the maneuver load factor n_z, one may derive the following parameters:

$$\frac{dBTL}{dn_z} = \frac{[W(CG - 0.25 + dC_M/dC_L)]}{x_t/c_w + dC_M/dC_L} \tag{2.22}$$

$$\frac{d\alpha_w}{dn_z} = \left(\frac{W - dBTL/dn_z}{q S_w C_{L\alpha w}}\right) \tag{2.23}$$

The tail load and wing angle of attack at zero load factor are derived using Eqs. (2.22) and (2.23):

$$BTL_o = (C_{Mo}qS_w)/(x_t/c_w + dC_M/dC_L) \qquad (2.24)$$

$$\alpha_{wo} = [-(BTL_o/qS_w) - C_{Lo}]/C_{L\alpha w} \qquad (2.25)$$

If the engine thrust moment arm z_e does not vary with load factor, then power-on conditions may be determined by modifying Eq. (2.24) to include the thrust term:

$$BTL_o = (C_{Mo}qS_w + T_{eng}z_e)/(x_t/c_w + dC_M/dC_L) \qquad (2.26)$$

If the aircraft configuration has engines located outboard on the wing, then the engine thrust moment arm will vary with wing deflection. Solution of power-on conditions may be obtained by using a variable moment arm in Eq. (2.26) to represent a given load factor or the effect may be included in Eq. (2.22) in which the moment arm varies directly with load factor.

2.3.4 Graphical Solution for Determining Balancing Tail Loads

A graphical solution for determining balancing tail loads, particularly usable when the tail-off aerodynamic coefficients are nonlinear, is shown in Fig. 2.6. The following relationships are assumed:

$$C_{La} = n_z W/qS_w \qquad (2.9)$$

$$C_L = C_{La} - BTL/qS_w \qquad (2.27)$$

$$BTL = [(CG - 0.25)C_{La} + C_{M0.25}]qS_w c_w/x_t \qquad (2.13)$$

The pitching moment coefficient $C_{M0.25}$ may be determined graphically as shown in Fig. 2.6 using the following procedure.

1) Establish the axis representing the center of gravity for the airplane condition, which for this example is noted as a forward center of gravity. This is shown in Fig. 2.6 as $(CG - 0.25)C_L$.

2) Calculate the airplane lift coefficient from Eq. (2.9).

3) Calculate the sloping line representing the tail-off lift coefficient as shown in Fig. 2.6.

4) The resulting intersection of this slope with the tail-off pitching moment curve will give the desired pitching moment coefficient $C_{M0.25}$ and the tail-off lift coefficient C_L.

5) The reference wing angle of attack is determined from the tail-off lift coefficients as shown in Fig. 2.6.

6) Balancing tail loads are then calculated from Eq. (2.13).

For power-on conditions the tail-off pitching moment curve must be corrected by the thrust increment as shown in Eq. (2.28):

$$\Delta C_M = T_{eng}z_e/(qS_w c_w) \qquad (2.28)$$

The graphical method of solving for balancing tail loads and wing angle of attack for nonlinear coefficients may also be solved on a personal computer by using the nonlinear representation of the lift and pitching moment curves in a table.

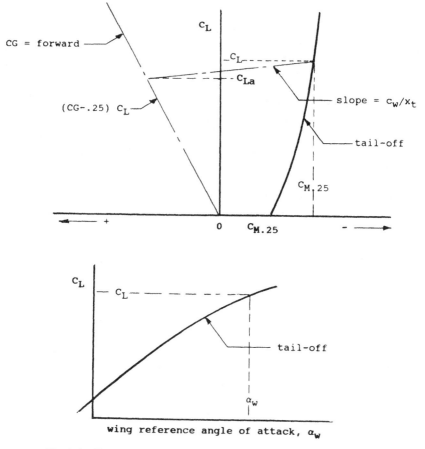

Fig. 2.6   Graphical solution for determining balancing tail loads.

2.3.5   Static Aeroelastic Effects

Static aeroelastic effects should be included in determining the tail-off lift and pitching moment characteristics of the airplane. NACA TN 3030[1] provides a method whereby wing loads may be determined for an elastic wing. Using this method for a steady-state symmetrical maneuver will provide wing loads, wing angle of attack, and balancing tail load for the condition under investigation. This same method can be used to determine the incremental lift and pitching moment coefficients due to aeroelasticity and so will allow correction of rigid data obtained from wind-tunnel model tests.

2.4   Abrupt Pitching Maneuvers

Abrupt pitching maneuvers when applied to structural load analysis are maneuvers involving a single rapid application of the elevator in a prescribed manner. These maneuvers are shown graphically in Fig. 2.7.

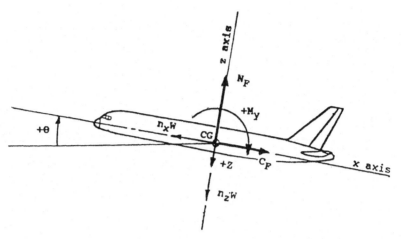

Fig. 2.7 **Forces acting on the airplane during abrupt pitching maneuvers; 1) pitching maneuvers are purely symmetrical and 2) airspeed and Mach number (hence altitude) are constant during maneuver.**

In general, two types of abrupt pitching maneuvers need to be considered: 1) abrupt unchecked elevator maneuver at V_A speed and 2) elevator-checked maneuver at V_A to V_D speeds.

2.4.1 Equations of Motion for Pitching Maneuvers

The equations of motion for pitching maneuvers are derived from the work of John Etkin in Ref. 2 and from Ref. 3, which was published in 1945 on the subject of pitch maneuver loads. References 4 and 5 are other historical sources for the equations of motion.

Making the assumption that the maneuvers are purely symmetrical and airspeed and altitude are held constant, one can derive the equations of motion representing translation along the z axis and rotation about the y axis:

$$\ddot{Z} = Z_\alpha \blacktriangle \alpha + Z_{\dot{\alpha}} \dot{\alpha} + Z_{\dot{\theta}} \dot{\theta} + Z_{\delta e} \delta_e \qquad (2.29)$$

$$\ddot{\theta} = M_\alpha \blacktriangle \alpha + M_{\dot{\alpha}} \dot{\alpha} + M_{\dot{\theta}} \dot{\theta} + M_{\delta e} \delta_e \qquad (2.30)$$

The relationship shown in Eq. (2.31) is obtained from Ref. 2:

$$\ddot{Z} = V(\dot{\alpha} - \dot{\theta}) \qquad (2.31)$$

Combining Eqs. (2.29–2.31), the equations of motion are as shown in Eqs. (2.32) and (2.33):

$$\dot{\alpha} = [Z_\alpha \blacktriangle \alpha + (V + Z_{\dot{\theta}}) \dot{\theta} + Z_{\delta e} \delta_e]/(V - Z_{\dot{\alpha}}) \qquad (2.32)$$

$$\ddot{\theta} = M_\alpha \blacktriangle \alpha + M_{\dot{\alpha}} \dot{\alpha} + M_{\dot{\theta}} \dot{\theta} + M_{\delta e} \delta_e \qquad (2.33)$$

Using the relationship from Eq. (2.31), one can derive the load factor along the z axis as shown in Eq. (2.34):

$$\blacktriangle n_z = -V(\dot{\alpha} - \dot{\theta})/g \qquad (2.34)$$

Table 2.5   Definition of stability derivatives used for pitch maneuver analyses

Stability derivatives:

$$Z_\alpha = -C_{L\alpha} q S_w / M \tag{2.32a}$$

$$Z_{\dot\alpha} = -L_{\alpha s} l_t \epsilon_\alpha / (MV) \tag{2.32b}$$

$$Z_{\dot\theta} = -L_{\alpha s} l_t / (MV) \tag{2.32c}$$

$$Z_{\delta e} = -L_{\delta e} / M \tag{2.32d}$$

$$M_\alpha = C_{M\alpha} q S_w c_w / I_y \tag{2.33a}$$

$$M_{\dot\alpha} = M_{\alpha s} l_t \epsilon_\alpha / (I_y V) \tag{2.33b}$$

$$M_{\dot\theta} = M_{\alpha s} l_t K_{wb} / (I_y V) \tag{2.33c}$$

K_{wb} = correction for wing–body effects due to pitching velocity $\dot\theta$

$$M_{\delta e} = M_{t\delta e} / I_y \tag{2.33d}$$

Tail-on static stability derivatives:

$$C_{L\alpha} = C_{L\alpha t o} + L_{\alpha s} (1 - \epsilon_\alpha) \tag{2.32e}$$

$$C_{M\alpha} = C_{M\alpha t o} + M_{\alpha s} (1 - \epsilon_\alpha) / (q S_w c_w) + C_{L\alpha} (CG - 0.25) \tag{2.33e}$$

Horizontal tail terms including body flexibility effects:

$$L_{\alpha s} = L_{t\alpha s} - l_t L_{t\alpha s} / K_{BB} \tag{2.32f}$$

$$M_{\alpha s} = (M_{t\alpha s} - l_t L_{t\alpha s}) / K_{BB} \tag{2.33b}$$

$$L_{\delta e} = [1 + K_{\delta e} L_{t\alpha s} / K_{BB} L_{t\delta e}] L_{t\delta e} \tag{2.32g}$$

$$M_{\delta e} = [1 + K_{\delta e} M_{t\alpha s} / K_{BB} M_{t\delta e}] M_{t\delta e} - l_t L_{\delta e} \tag{2.33g}$$

Body flexibility parameters:

$$K_{BB} = 1 - \left(\frac{d\alpha_s}{dL_t}\right) L_{t\alpha s} - \left(\frac{d\alpha_s}{dM_t}\right) M_{t\alpha s} \tag{2.32h}$$

$$K_{\delta e} = \left(\frac{d\alpha_s}{dL_t}\right) L_{t\delta e} - \left(\frac{d\alpha_s}{dM_t}\right) M_{t\delta e} \tag{2.33h}$$

Equations (2.32) and (2.33) may now be solved on a personal computer using finite difference techniques or other methods to determine the time history of the airplane load factor and related parameters such as pitch velocity, pitch acceleration, and wing angle of attack.

The stability derivatives in Eqs. (2.32) and (2.33) are defined in Table 2.5.

2.5   Abrupt Unchecked Pitch Maneuvers

Per the requirements of FAR 25.331(c), an abrupt unchecked pitch maneuver analysis is made at V_A speeds, in which the airplane is assumed to be flying in a steady level flight, and except as limited by pilot effort in accordance with FAR 25.397(b), the pitching control is suddenly moved to obtain extreme positive pitching acceleration (nose up).

The dynamic response or, at the option of the applicant, the transient rigid-body response of the airplane must be taken into account in determining the tail load. Airplane loads that occur after the normal acceleration at the center of gravity

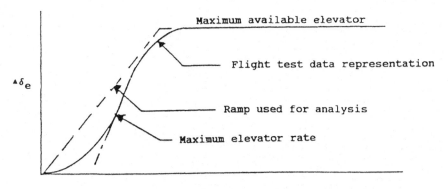

Fig. 2.8 Abrupt elevator motion analysis ramp vs flight test data.

exceeding the maximum positive limit maneuvering load factor n_z need not be considered [see FAR 25.331(c)(1)].

The question must be raised as to the shape of the elevator motion during this abrupt maneuver. From flight test data for several types of jet transports, the motion seems to follow what may be called an "S" curve; i.e., the elevator motion with time takes the shape of an elongated "S" starting slowly near $t = 0$, then reaching the maximum system rate capability at $\Delta\delta_{max}/2$, and then slowing as the maximum elevator angle is reached. This is typified in Fig. 2.8.

2.5.1 Abrupt Unchecked Elevator Motion

An example of the abrupt elevator time histories is shown in Fig. 2.9 for two types of input: 1) a linear ramp elevator using a rate of application such that the

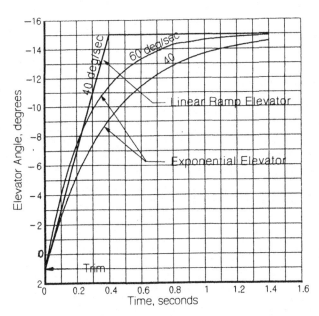

Fig. 2.9 Elevator time histories used for abrupt unchecked maneuver analysis.

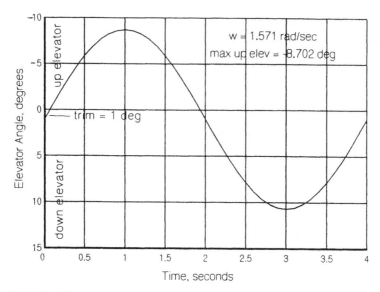

Fig. 2.10   Sinusoidal elevator input used for checked maneuver analysis.

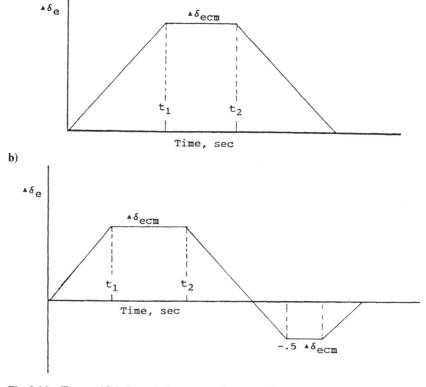

Fig. 2.11   Trapezoidal-shaped elevator motion per military requirements: a) forward center of gravity position and b) aft center of gravity position.

motion encompasses that which would be measured in flight test and 2) an expo-
nentially shaped elevator input that is based on British Civil Aviation Authority
(CAA) requirements for airplanes certified during the early 1980s.

The example time history solutions of Eqs. (2.32) and (2.33) for an abrupt
unchecked elevator maneuver, shown in Tables 2.6 and 2.7, use an integration
time increment of 0.01 s. Tables 2.6 show an abrupt-up elevator with a 40-deg/s
linear ramp. Tables 2.7 show an abrupt-up elevator, with an exponential input
at 60-deg/s maximum rate (this rate approaches the maximum capability of the
power control units for the aircraft shown).

Horizontal tail loads are shown in these tables as the critical load parameter for
this maneuver. The analysis ends after the design maneuver load factor is exceeded
per the criteria stated in FAR 25.331(c)(1).

2.5.2 Maximum Available Up Elevator

The maximum available up elevator required for the unchecked maneuver is
limited by elevator stops, 300-lb pilot effort, or by the power control unit capability
per FAR 25.331(c)(1). The maximum available elevator is determined using the
methodology discussed in Sec. 12.2 of Chapter 12.

Table 2.6a **Abrupt-up unchecked pitch maneuver analysis
using linear ramp inputs (airplane response)**

Cond.: 2	Alt., ft = 0	V_e, keas = 282		Mach = 0.426	
Gross weight, lb = 254,500		CG, % mac/100 = 0.202		I_y, E6 slug ft^2 = 7.3	
I	Time, s	δ_e, deg	n_z	α_w, deg	$\ddot{\theta}$, rad/s^2	$\dot{\theta}$, rad/s
		Time history analysis ($\Delta t = 0.01$ s)				
0	0.00	1.000	1.000	4.441	0.00000	0.00000
10	0.10	−3.000	0.946	4.471	0.14189	0.00791
20	0.20	−7.000	0.914	4.606	0.27072	0.02929
30	0.30	−11.000	0.914	4.909	0.38577	0.06281
40	0.40	−14.990	0.955	5.429	0.48633	0.10705
50	0.50	−14.990	1.099	6.177	0.43123	0.15267
60	0.60	−14.990	1.276	7.105	0.37529	0.19271
70	0.70	−14.990	1.477	8.172	0.31972	0.22718
80	0.80	−14.990	1.696	9.337	0.26554	0.25615
90	0.90	−14.990	1.925	10.564	0.21362	0.27983
100	1.00	−14.990	2.159	11.820	0.16466	0.29847
110	1.10	−14.990	2.393	13.078	0.11922	0.31241
120	1.20	−14.990	2.622	14.312	0.07769	0.32201
		Time of maximum horizontal tail load				
40	0.40	−14.990	0.955	5.429	0.48633	0.10705
		Time of maximum horizontal tail pitching moment				
40	0.40	−14.990	0.955	5.429	0.48633	0.10705

**Table 2.6b Abrupt-up unchecked pitch maneuver analysis
using linear ramp inputs (horizontal tail loads)**

Cond.: 2	Alt., ft $= 0$	V_e, keas $= 282$	Mach $= 0.426$
Gross weight, lb $= 254{,}500$	CG, % mac/100 $= 0.202$	I_y, E6 slug ft$^2 = 7.3$	

I	Time, s	n_z	δ_e, deg	α_s, deg	L_t, lb limit	M_t, 10^{-6} in.-lb limit
\multicolumn{7}{c}{*Time history analysis* ($\Delta t = 0.01$ s)}						
0	0.00	1.000	1.000	1.054	$-6{,}360$	-0.194
10	0.10	0.946	-3.000	1.293	$-21{,}268$	0.564
20	0.20	0.914	-7.000	1.736	$-34{,}635$	1.299
30	0.30	0.914	-11.000	2.402	$-46{,}317$	2.010
40	0.40	0.955	-14.990	3.304	$-56{,}181$	2.693
50	0.50	1.099	-14.990	4.209	$-49{,}358$	2.593
60	0.60	1.276	-14.990	5.151	$-42{,}252$	2.489
70	0.70	1.477	-14.990	6.108	$-35{,}039$	2.384
80	0.80	1.696	-14.990	7.058	$-27{,}875$	2.280
90	0.90	1.925	-14.990	7.984	$-20{,}893$	2.178
100	1.00	2.159	-14.990	8.871	$-14{,}204$	2.080
110	1.10	2.393	-14.990	9.707	$-7{,}898$	1.988
120	1.20	2.622	-14.990	10.483	$-2{,}046$	1.903
\multicolumn{7}{c}{*Time of maximum horizontal tail load*}						
40	0.40	0.955	-14.990	3.304	$-56{,}181$	2.693
\multicolumn{7}{c}{*Time of maximum horizontal tail pitching moment*}						
40	0.40	0.955	-14.990	3.304	$-56{,}181$	2.693

2.6 Abrupt Checked Maneuvers (Commercial Requirements)

A checked maneuver, based on a rational pitching control motion vs time profile, must be established in which the design limit load factor specified in FAR 25.337 will not be exceeded. Unless lesser values cannot be exceeded, the airplane response must result in pitching accelerations not less than specified in Eqs. (2.37) and (2.38) as discussed in Secs. 2.8.1 and 2.8.2.

Per JAR 25.331(c)(2), checked pitch maneuvers must be analyzed for nose-up conditions to the maximum positive design limit load factor and for nose-down conditions to a load factor of zero.

The airplane is assumed to be flying in steady flight at any speed between V_A and V_D when the cockpit pitch control is moved rapidly in a sinusoidal motion as shown in Fig. 2.10. The elevator is moved rapidly in one direction, then reversed to a position well beyond the original trim position before returning to the trim position.

The elevator motion is defined in Eq. (2.35):

$$\delta_e = \delta_{e0} \sin wt \tag{2.35}$$

Table 2.7a Abrupt-up unchecked pitch maneuver analysis using exponential shape inputs (airplane response)

Max. elevator rate = 60 deg/s

Cond.: 2   Alt., ft = 0 V_e, keas = 282   Mach = 0.426

Gross weight, lb = 254,500 CG, % mac/100 = 0.202 I_y, E6 slug ft^2 = 7.3

Max. elev., deg = −14.990

I	Time, s	δ_e, deg	n_z	α_w, deg	$\ddot{\theta}$, rad/s^2	$\dot{\theta}$, rad/s
			Time history analysis ($\Delta t = 0.01$ s)			
0	0.00	1.000	1.000	4.441	0.00000	0.00000
10	0.10	−4.003	0.933	4.481	0.17698	0.01044
20	0.20	−7.440	0.915	4.643	0.28221	0.03441
30	0.30	−9.802	0.943	4.966	0.33727	0.06601
40	0.40	−11.425	1.014	5.462	0.35742	0.10108
50	0.50	−12.541	1.125	6.126	0.35357	0.13677
60	0.60	−13.307	1.268	6.941	0.33354	0.17113
70	0.70	−13.834	1.438	7.884	0.30302	0.20288
80	0.80	−14.195	1.629	8.928	0.26616	0.23119
90	0.90	−14.444	1.834	10.044	0.22598	0.25561
100	1.00	−14.615	2.049	11.207	0.18473	0.27594
110	1.10	−14.732	2.267	12.388	0.14403	0.29217
120	1.20	−14.813	2.485	13.566	0.10509	0.30441
130	1.30	−14.868	2.698	14.718	0.06873	0.31290
			Time of maximum horizontal tail load			
40	0.40	−11.425	1.014	5.462	0.35742	0.10108
			Time of maximum horizontal tail pitching moment			
67	0.67	−13.696	1.385	7.589	0.31301	0.19368

where δ_e is the elevator angle, δ_{e0} is the elevator required to obtain the required design limit load factor n_z, and w is the control surface rate equal to the undamped natural frequency of the short period rigid mode in pitch but not less than $(\pi V_e)/(2V_A)$ where V_e and V_A are airspeeds in knots equivalent airspeed (keas), radians per second.

The short period rigid mode in pitch may be approximated from Eq. (2.36),[2] using the derivatives defined in Table 2.5:

$$w_n = [(Z_\alpha M_{\dot{\theta}}/V) - M_\alpha]^{\frac{1}{2}} \qquad \text{(rad/s)} \qquad (2.36)$$

Solution of the equations of motion, Eqs. (2.32) and (2.33), using the sinusoidal elevator input defined by Eq. (2.35) may now be accomplished.

An example of an elevator-checked maneuver analysis is shown in Tables 2.8 using the equations of motion and elevator angle defined earlier. The elevator angle δ_{e0} is obtained by iteration such that the required limit load factor is achieved. For the example shown, the integration time increment is 0.01 s.

**Table 2.7b   Abrupt-up unchecked pitch maneuver analysis
using exponential shape inputs (horizontal tail loads)**

Max. elevator rate = 60 deg/s

Cond.: 2	Alt., ft = 0		V_e, keas = 282		Mach = 0.426

Gross weight, lb = 254,500 $\quad$ CG, % mac/100 = 0.202 $\quad$ I_y, E6 slug ft^2 = 7.3

I	Time, s	n_z	δ_e, deg	α_s, deg	L_t, lb limit	M_t, 10^{-6} in.-lb limit
			Time history analysis ($\Delta t = 0.01$ s)			
0	0.00	1.000	1.000	1.054	−6,360	−0.194
10	0.10	0.933	−4.003	1.360	−24,950	0.753
20	0.20	0.915	−7.440	1.827	−35,789	1.375
30	0.30	0.943	−9.802	2.428	−41,124	1.772
40	0.40	1.014	−11.425	3.136	−42,564	2.012
50	0.50	1.125	−12.541	3.924	−41,274	2.144
60	0.60	1.268	−13.307	4.769	−38,105	2.202
70	0.70	1.438	−13.834	5.647	−33,688	2.208
80	0.80	1.629	−14.195	6.536	−28,493	2.181
90	0.90	1.834	−14.444	7.418	−22,877	2.133
100	1.00	2.049	−14.615	8.278	−17,105	2.072
110	1.10	2.267	−14.732	9.102	−11, 383	2.004
120	1.20	2.485	−14.813	9.879	−5,861	1.935
130	1.30	2.698	−14.868	10.600	−654	1.866
			Time of maximum horizontal tail load			
40	0.40	1.014	−11.425	3.136	−42564	2.012
			Time of maximum horizontal tail pitching moment			
67	0.67	1.385	−13.696	5.381	−35113	2.210

Per the requirements of JAR 25.331(c)(2), it is necessary to analyze only three-quarters of the cyclic movement of the cockpit pitch control, assuming that the final return to the original trim position is achieved in a less sudden manner.

A comparison of the time history pitching accelerations with the requirements of FAR 25.331(c)(2) is shown in Table 2.9.

2.7   Abrupt Checked Maneuvers (Military Requirements)

A word must be said about other elevator time history shapes. The military specifications require a trapezoidal-shaped elevator input that is checked back to neutral for conditions with forward center of gravity positions and checked back beyond neutral (or the trim point) to 50% of the original input elevator as shown in Fig. 2.11 for aft center of gravity positions.

The solution to the trapezoidal-shaped elevator requires an iteration process to determine $\Delta\delta_{cm}$, t_1, and t_2 to obtain the desired maneuver load factor. The slopes are generally selected as the maximum elevator rates available. The solution may be accomplished with computers, but it is time consuming.

Table 2.8a Elevator checked maneuver analysis using sinusoidal inputs (airplane response)

Max. n_z obtained = 2.500 at T = 2.00 s

Cond.: 2   Alt., ft = 0 V_e, keas = 282   Mach = 0.426

Gross weight, lb = 254,500 CG, % mac/100 = 0.202 I_y, E6 slug ft^2 = 7.3

Max. elev., deg = −14.990

I	Time, s	δ_e, deg	n_z	α_w, deg	$\ddot{\theta}$, rad/s^2	$\dot{\theta}$, rad/s
			Time history analysis (Δt = 0.01 s)			
0	0.00	1.000	1.000	4.441	0.00000	0.00000
10	0.10	−0.518	0.980	4.452	0.05383	0.00301
20	0.20	−1.998	0.968	4.504	0.10137	0.01106
30	0.30	−3.405	0.969	4.617	0.14118	0.02346
40	0.40	−4.703	0.987	4.809	0.17214	0.03936
50	0.50	−5.860	1.024	5.089	0.19347	0.05782
60	0.60	−6.849	1.080	5.462	0.20476	0.07788
70	0.70	−7.644	1.157	5.927	0.20590	0.09850
80	0.80	−8.227	1.252	6.475	0.19713	0.11869
90	0.90	−8.582	1.363	7.094	0.17899	0.13748
100	1.00	−8.702	1.488	7.769	0.15227	0.15397
110	1.10	−8.582	1.622	8.478	0.11804	0.16737
120	1.20	−8.227	1.762	9.198	0.07756	0.17700
130	1.30	−7.644	1.902	9.907	0.03223	0.18229
140	1.40	−6.849	2.038	10.577	−0.01640	0.18286
150	1.50	−5.860	2.164	11.185	−0.06673	0.17846
160	1.60	−4.703	2.276	11.705	−0.11711	0.16901
170	1.70	−3.405	2.370	12.116	−0.16593	0.15460
180	1.80	−1.998	2.440	12.397	−0.21164	0.13546
190	1.90	−0.518	2.485	12.534	−0.25279	0.11199
200	2.00	1.000	2.500	12.512	−0.28809	0.08471
210	2.10	2.518	2.484	12.325	−0.31640	0.05429
220	2.20	3.998	2.437	11.968	−0.33680	0.02145
230	2.30	5.405	2.356	11.443	−0.34861	−0.01295
240	2.40	6.703	2.245	10.756	−0.35136	−0.04804
250	2.50	7.860	2.103	9.918	−0.34488	−0.08289
260	2.60	8.849	1.934	8.944	−0.32921	−0.11659
270	2.70	9.644	1.741	7.854	−0.30467	−0.14824
280	2.80	10.227	1.528	6.671	−0.27182	−0.17696
290	2.90	10.582	1.299	5.420	−0.23144	−0.20198

Table 2.8a   Elevator checked maneuver analysis using sinusoidal inputs (airplane response) (continued)

I	Time, s	δ_e, deg	n_z	α_w, deg	$\ddot{\theta}$, rad/s²	$\dot{\theta}$, rad/s
		Time history analysis ($\Delta t = 0.01$ s)				
300	3.00	10.702	1.060	4.131	−0.18453	−0.22259
310	3.10	10.582	0.817	2.832	−0.13224	−0.23821
320	3.20	10.227	0.575	1.555	−0.07590	−0.24836
330	3.30	9.644	0.339	0.331	−0.01692	−0.25272
		Time of maximum horizontal tail load				
230	2.30	5.405	2.356	11.443	−0.34861	−0.01295
61	0.61	−6.938	1.087	5.505	0.20533	0.07993
		Time of maximum pitch acceleration				
66	0.66	−7.351	1.124	5.731	0.20665	0.9025
238	2.38	6.453	2.270	10.906	−0.35155	−0.04101

2.8   Minimum Pitch Acceleration Requirements

The selected elevator time profile must result in pitching accelerations not less than those specified in FAR 25.331(c)(2), unless lesser values cannot be exceeded due to the design configuration under investigation.

For certain aircraft configurations such as the very large wide-body jets, a rational analysis using conservative elevator time profiles may produce pitching accelerations less than the minimum required per FAR 25.331(c)(2). For smaller to medium-sized commercial transports, a rational analysis may result in pitching accelerations larger than the minimum required per FAR 25.331(c)(2).

The airspeeds and load factors shown in Eqs. (2.37) and (2.38) are defined as follows: n_z is the positive design load factor at the airspeed under consideration (see Table 2.1) and V_e is the equivalent airspeed (keas).

2.8.1   Positive Pitching Acceleration

A positive pitching acceleration (nose up) is assumed to be reached concurrently with the airplane load factor of 1.0 [points A_1 to D_1 of FAR 25.333(b)]. The positive acceleration must be equal to at least

$$\ddot{\theta} = 39 n_z (n_z - 1.5)/V_e \qquad \text{(rad/s}^2) \qquad (2.37)$$

2.8.2   Negative Pitching Acceleration

A negative pitching acceleration (nose down) is assumed to be reached concurrently with the positive maneuvering load factor [points A_2 to D_2 of FAR 25.333(b)]. The negative pitching acceleration must be equal to at least

$$\ddot{\theta} = -26 n_z (n_z - 1.5)/V_e \qquad \text{(rad/s}^2) \qquad (2.38)$$

Table 2.8b Elevator checked maneuver analysis using sinusoidal inputs (horizontal tail loads)

Max. n_z obtained $= 2.500$ at $T = 2.00$ s

Cond.: 2   Alt., ft $= 0$ V_e, keas $= 282$ Mach $= 0.426$

Gross weight, lb $= 254{,}500$ CG, % mac/100 $= 0.202$ I_y, E6 slug ft$^2 = 7.3$

Max. elev., deg $= -14.990$

I	Time, s	n_z	δ_e, deg	α_s, deg	L_t, lb limit	M_t, 10^{-6} in.-lb limit
			Time history analysis ($\Delta t = 0.01$ s)			
0	0.00	1.000	1.000	1.054	$-6{,}360$	-0.194
10	0.10	0.980	-0.518	1.145	$-12{,}016$	0.094
20	0.20	0.968	-1.998	1.311	$-16{,}947$	0.365
30	0.30	0.969	-3.405	1.555	$-20{,}976$	0.614
40	0.40	0.987	-4.703	1.878	$-23{,}969$	0.833
50	0.50	1.024	-5.860	2.271	$-25{,}836$	1.017
60	0.60	1.080	-6.849	2.727	$-26{,}530$	1.160
70	0.70	1.157	-7.644	3.232	$-26{,}045$	1.261
80	0.80	1.252	-8.227	3.770	$-24{,}419$	1.316
90	0.90	1.363	-8.582	4.324	$-21{,}724$	1.324
100	1.00	1.488	-8.702	4.875	$-18{,}068$	1.287
110	1.10	1.622	-8.582	5.403	$-13{,}589$	1.206
120	1.20	1.762	-8.227	5.888	$-8{,}446$	1.083
130	1.30	1.902	-7.644	6.311	$-2{,}821$	0.922
140	1.40	2.038	-6.849	6.655	$3{,}094$	0.728
150	1.50	2.164	-5.860	6.904	$9{,}099$	0.507
160	1.60	2.276	-4.703	7.044	$14{,}993$	0.265
170	1.70	2.370	-3.405	7.066	$20{,}579$	0.008
180	1.80	2.440	-1.998	6.962	$25{,}673$	-0.256
190	1.90	2.485	-0.518	6.730	$30{,}102$	-0.521
200	2.00	2.500	1.000	6.369	$33{,}718$	-0.779
210	2.10	2.484	2.518	5.883	$36{,}393$	-1.023
220	2.20	2.437	3.998	5.280	$38{,}029$	-1.246
230	2.30	2.356	5.405	4.571	$38{,}557$	-1.444
240	2.40	2.245	6.703	3.770	$37{,}939$	-1.610
250	2.50	2.103	7.860	2.894	$36{,}170$	-1.741
260	2.60	1.934	8.849	1.963	$33{,}278$	-1.832
270	2.70	1.741	9.644	0.997	$29{,}320$	-1.882
280	2.80	1.528	10.227	0.020	$24{,}385$	-1.889
290	2.90	1.299	10.582	-0.945	$18{,}588$	-1.852

Table 2.8b   Elevator checked maneuver analysis using sinusoidal inputs (horizontal tail loads) (continued)

I	Time, s	n_z	δ_e, deg	α_s, deg	L_t, lb limit	M_t, 10^{-6} in.-lb limit
			Time history analysis ($\Delta t = 0.01$ s)			
300	3.00	1.060	10.702	−1.876	12,070	−1.773
310	3.10	0.817	10.582	−2.749	4,990	−1.654
320	3.20	0.575	10.227	−3.542	−2,477	−1.497
330	3.30	0.339	9.644	−4.236	−10,142	−1.306
			Time of maximum horizontal tail load			
230	2.30	2.356	5.405	4.571	38,557	−1.444
61	0.61	1.087	−6.938	2.775	−26,534	1.172
			Time of maximum pitch acceleration			
66	0.66	1.124	−7.351	3.025	−26,379	1.226
238	2.38	2.270	6.453	3.936	38,155	−1.580

Table 2.9   Comparison of pitching accelerations from time history analysis with FAR 25.331(c)(2)

	Pitching acceleration, rad/s^2		
	Abrupt-up elevator linear ramp	Abrupt-up elevator exponential analysis	Checked maneuver
Time history analysis	0.486[b]	0.357[c]	−0.352[d]
Minimum requirement,[a] FAR 25.331(c)(2)	0.346	0.346	−0.231

[a] Where $n_z = 2.5$ and $v_e = 282$ keas.
[b] See Table 2.6a.
[c] See Table 2.7a.
[d] See Table 2.8a.

References

[1]Gray, W. L., and Schenk, K. M., "A Method for Calculating the Subsonic Steady-State Loading on an Airplane Wing of Arbitrary Plan Form and Stiffness," NACA TN 3030, Dec. 1953.

[2]Etkin, B., *Dynamics of Flight Stability and Control,* Wiley, New York, 1959.

[3]Kelly, J., and Missall, J. W., "Maneuvering Horizontal Tail Loads," U.S. Army Air Forces TR 5185, Wright Field, Dayton, OH, Jan. 1945.

[4]Pearson, H. A., "Derivation of Charts for Determining the Horizontal Tail Load Variation with Any Elevator Motion," NACA Rept. 759, Nov. 1942.

[5]Pearson, H. A., McGowan, W. A., and Donegan, J. J., "Horizontal Tail Loads in Maneuvering Flight," NACA Rept. 1007, Feb. 1950.

3
Rolling Maneuvers

Rolling maneuvers are unsymmetrical maneuvers involving application of the lateral control devices to produce motion of the airplane about the x axis. Rolling maneuvers are accomplished in conjunction with a specified symmetrical load factor.

The following assumptions are made for analytical purposes. 1) Airspeed and Mach number (hence altitude) are assumed constant throughout the rolling maneuver. 2) For structural load analyses designed to commercial transport regulations, FAR 25.349, cross-coupling effects between the yaw and roll degrees of freedom can be neglected.

3.1 Parameters Required for Structural Load Analyses

The parameters required for structural load analyses for rolling maneuver conditions are shown in Table 3.1. The roll accelerations and velocities, necessary for calculation of loads on the wing and empennage due to rolling maneuvers, may be determined using the methods developed in this chapter.

Asymmetrical conditions are defined as the incremental loads due to roll or yaw before the inclusion of the symmetrical flight load increments.

3.2 Symmetrical Load Factors for Rolling Maneuvers

Except where limited by maximum static lift coefficients for the airplane, symmetrical load factors for rolling maneuver conditions are shown in Table 3.2, per FAR 25.349(a).

The initial conditions for rolling maneuvers at positive design maneuver factors are fairly logical in that the airplane may be rolled from a turn in one direction to a turn in the other direction, maintaining a constant load factor during the maneuver. These maneuvers may be easily accomplished during flight tests.

Rolling maneuvers at a load factor of zero are not necessarily logical but are done analytically to give bounds to the design problem.

3.2.1 Initial Bank Angle as a Function of Load Factor

The airplane bank angle during a wind-up turn is a direct function of the airplane load factor attained during the turn. The relationship between load factor and bank angle may be derived from the forces shown in Fig. 3.1:

$$n_z W \cos \phi_0 = W \qquad (3.1)$$

$$\phi_0 = \cos^{-1}(1/n_z) \qquad (3.2)$$

where n_z is the airplane load factor (g), W is the airplane gross weight (lb), and ϕ_0 is the airplane initial bank angle (rad).

The variation of load factor n_z with bank angle is shown in Fig. 3.2 for a coordinated turn.

Table 3.1   Unsymmetrical load analysis parameters required

Unsymmetrical condition	Symmetrical condition load factor	Asymmetrical condition
Roll maneuvers	See Table 3.2	Roll rates and accelerations $\dot{\phi}$ and $\ddot{\phi}$

Table 3.2   Symmetrical load factors
for roll maneuvers

Maneuvers	Symmetrical load factor
Positive	$(2/3)n_z$ shown in Tables 2.1 and 2.2
Negative	0

a)

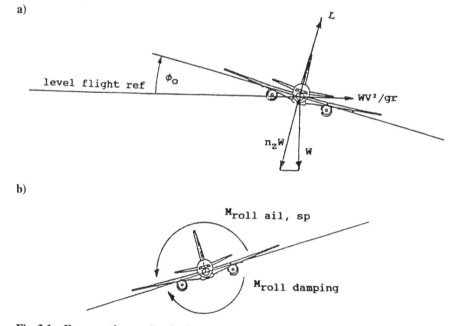

b)

Fig. 3.1   Forces acting on the airplane during a roll maneuver: a) initial condition at $t = 0$ and b) condition during roll maneuver.

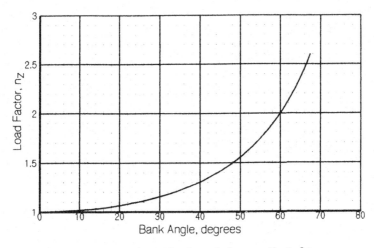

Fig. 3.2   Load factor vs bank angle for coordinated turn.

3.3   Control Surface Deflections for Rolling Maneuvers

The requirements of FAR 25.349(a) stipulate application of lateral control (ailerons plus spoilers where applicable) as a function of structural design airspeeds (see Table 3.3).

3.4   Equations of Motion for Rolling Maneuvers

A complete set of the equations of motion for an airplane involving side translation, yaw, and roll degrees of freedom is derived in Ref. 1.

Using the assumptions previously stated in this chapter, and given that the angle of attack effects during the rolling maneuver are based on the initial condition and are assumed unchanged during the maneuver, the simplified equation of motion due to roll may be determined:

$$\left(I_x - M_{\ddot{\phi}}\right)\ddot{\phi} + M_{\dot{\phi}}\dot{\phi} = M_\delta \delta \tag{3.3}$$

where I_x is the airplane moment of inertia in roll (slug ft²), $M_{\ddot{\phi}}$ is the rolling moment due to aeroelasticity for a roll acceleration of 1.0 rad/s² (ft-lb), $M_{\ddot{\phi}}$ is zero for a rigid wing analysis, $M_{\dot{\phi}}$ is the roll damping moment for rolling velocity = 1.0 rad/s (ft-lb), and $M_\delta \delta$ is the rolling moment due to lateral control application (including ailerons plus spoilers) (ft-lb).

Table 3.3   Lateral control deflections required

Airspeeds	Lateral control applied
V_A	Sudden deflection to maximum available
V_C	$\dot{\phi}_{max} = \dot{\phi}_{max}$ at V_A
V_D	$\dot{\phi}_{max} = \left(\frac{1}{3}\right)\dot{\phi}_{max}$ at V_A

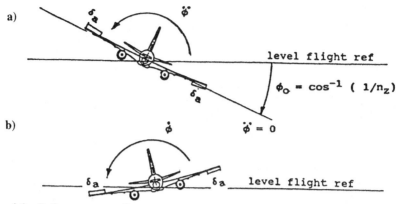

Fig. 3.3   Roll maneuvers for commercial transport aircraft (roll termination not shown): a) roll initiation starting from a wind-up turn and b) steady roll condition.

The applied rolling moment due to aileron and spoilers, assuming a linear analysis, may be written as

$$M_\delta \delta = M_{ail}\delta_{ail} + M_{sp}\delta_{sp} \tag{3.4}$$

where M_{ail} is the rolling moment due to unit aileron deflection (ft-lb/deg), and M_{sp} is the rolling moment due to unit spoiler deflection (ft-lb/deg).

3.5   Maximum Rolling Acceleration and Velocity Criteria

FAR 25.349(a)(1) requires that two conditions be investigated as noted in Fig. 3.3: 1) the maximum steady rolling velocity condition, which by definition will have zero rolling acceleration, and 2) conditions corresponding to maximum angular accelerations, which must be investigated for aircraft with engines or other weight concentrations outboard of the fuselage. The rolling velocity may be assumed zero in the absence of a rational time history investigation of the maneuver.

3.5.1   Maximum Steady Roll Condition

By combining Eqs. (3.3) and (3.4), the maximum steady roll condition may be calculated by assuming zero roll acceleration:

$$\dot{\phi}_{max} = (M_{ail}\delta_{ail} + M_{sp}\delta_{sp})/M_{\dot{\phi}} \tag{3.5}$$

3.5.2   Maximum Rolling Acceleration Condition

In a similar manner, if the assumption is made in Eq. (3.3) that the airplane roll velocity is zero, then the maximum roll acceleration may be determined:

$$\ddot{\phi}_{max} = (M_{ail}\delta_{ail} + M_{sp}\delta_{sp})/(I_x - M_{\ddot{\phi}}) \tag{3.6}$$

This condition, known as roll initiation, is conservative; hence the regulations allow a rational time history analysis for determining design roll accelerations.

Table 3.4a   Exact solution to equation of motion defined in Eq. (3.3) for rolling maneuvers assuming linear control surface motion (equations of motion)

Let

$$p = \dot{\phi} \tag{3.3a}$$
$$\dot{p} = \ddot{\phi} \tag{3.3b}$$
$$k_1 = M_{\dot{\phi}} / \left(I_x - M_{\ddot{\phi}} \right) \tag{3.3c}$$
$$k_2 = M_\delta / \left(I_x - M_{\dot{\phi}} \right) \tag{3.3d}$$

then Eq. (3.3) can be written as

$$\dot{p} + k_1 p = k_2 \delta \tag{3.3e}$$

If the applied rolling moment is assumed linear with time as shown in Fig. 4.4, then Eq. (3.3e) becomes, for $t < t_1$,

$$\dot{p} + k_1 p = k_3 t \tag{3.3f}$$

where

$$k_3 = k_2 \delta_{\max} / t_1 \tag{3.3g}$$

and for $t_1 < t < t_2$,

$$\dot{p} + k_1 p = k_4 \tag{3.3h}$$

where

$$k_4 = k_2 \delta_{\max} \tag{3.3i}$$

and for $t_2 < t < t_3$,

$$\dot{p} + k_1 p = k_5 (t - t_3) \tag{3.3j}$$

where

$$k_5 = k_2 \delta_{\max} (t_2 - t_3) \tag{3.3k}$$

3.5.3   Rational Time History Investigation

The maximum rolling acceleration and velocity may be obtained from solution of the equation of motion, Eq. (3.3). Two solutions are considered as follows:

1) By defining the lateral control inputs as a linear function with time as shown in Fig. 3.4, an exact solution may be derived as shown in Tables 3.4. An analysis is shown in Table 3.5 for the exact solution using a linear lateral control input.

The relationship of lateral control wheel position vs aileron and spoiler angles shown in Fig. 3.5 indicates the lateral control inputs are not linear with respect to resulting spoiler angles.

2) An approximate solution using finite difference techniques[2,3] to solve the equation of motion may also be used. An analysis is shown in Table 3.6 assuming a linear lateral control input and an integration time increment of 0.02 s.

Since the differences between the results of these two analyses are small, the approximate method could be used for solution of nonlinear lateral control inputs by defining the input vs time in a table for solution on a personal computer.

The maximum rolling acceleration and velocity defined by the simplified analysis represented by Eqs. (3.5) and (3.6) are shown in Tables 3.5 and 3.6. As can be seen in these examples, the rolling velocity has almost attained the steady-state condition as calculated using Eq. (3.5).

Comparison of the maximum rolling acceleration computed using Eq. (3.6) with the time history results indicates that the simplified method gives very conservative accelerations when compared with a rational time history.

Table 3.4b   Exact solution to equation of motion for rolling maneuvers assuming linear control surface motion (solution to equations of motion)

For $t < t_1$:

$$\blacktriangle\phi = k_3\left[k_1 t^2/2 - t - e^{-k_1 t}/k_1 + 1/k_1\right]/(k_1)^2 \qquad (3.3l)$$

$$\dot\phi = k_3\left[k_1 t - 1 + e^{-k_1 t}\right]/(k_1)^2 \qquad (3.3m)$$

$$\ddot\phi = k_3\left[1 - e^{-k_1 t}\right]/k_1 \qquad (3.3n)$$

For $t_1 < t < t_2$:

$$\blacktriangle\phi = k_4 t/k_1 - k_3\left[1 - e^{+k_1 t}/k_1\right]e^{-k_1 t}/k_1^3 + C \qquad (3.3o)$$

$$\dot\phi = k_4/k_1 + k_3[(1 - e^{+k_1 t})e^{-k_1 t}]/k_1^2 \qquad (3.3p)$$

$$\ddot\phi = -k_3(1 - e^{+k_1 t})e^{-k_1 t}/k_1 \qquad (3.3q)$$

where

$$C = \phi(t_1) - k_4 t_1/k_1 + k_3(1 - e^{+k_1 t})e^{-k_1 t}/k_1^3 \qquad (3.3r)$$

For $t_2 < t < t_3$:

$$\blacktriangle\phi = k_5\left(t^2/2k_1 - t/k_1^2\right) - k_5 t_3 t/k_1 - Ce^{-k_1 t}/k_1 + D \qquad (3.3s)$$

$$\dot\phi = k_5\left(t/k_1 - 1/k_1^2\right) - k_5 t_3/k_1 + Ce^{-k_1 t} \qquad (3.3t)$$

$$\ddot\phi = k_5/k_1 - k_1 Ce^{-k_1 t} \qquad (3.3u)$$

where

$$D = \phi(t_2) - k_5\left[t_2^2/(2k_1) - t/k_1^2\right] - k_5 t_3 t_2/k_1 + Ce^{-k_1 t_2}/k_1 \qquad (3.3v)$$

As noted in Fig. 3.5 the aileron angle is linear with wheel input, but because of the delay in spoiler motion with wheel angle, the resulting motion is nonlinear. Thus the lateral control inputs are not linear as assumed for this example. Since design rolling maneuver conditions are for maximum wheel input, hence maximum aileron and spoiler angles (except as limited by blowdown limits), the error in assuming linear inputs will produce conservative roll rates and accelerations.

Equations (3.3f), (3.3h), and (3.3j) are differential equations of the first order and may be solved using integrating factors as shown in Ref. 2. The solutions shown in Eqs. (3.3l–3.3u) may be derived.

The variation of maximum acceleration with input time t_1 is shown in Fig. 3.6 for the same flight condition shown in Tables 3.5 and 3.6. A significant reduction in rolling acceleration is apparent if the maximum input time t_1 of 0.40 s is assumed.

Lateral control maximum input rates are calculated in Table 3.7 as a function of time t_1. In the example shown, the value used for this analysis is based on the

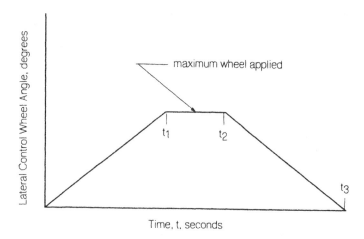

Fig. 3.4   Lateral control motion for roll maneuver.

Table 3.5 Rolling maneuver analysis using exact solution shown in Table 3.4b with linear lateral control surface motion defined in Fig. 3.4

$t_1, s = 0.4$ $t_2, s = 2.4$ $t_3, s = 2.8$

Time, s	Wheel, deg	Roll angle, deg	Rolling velocity, deg/s	Rolling acceleration, rad/s^2
0.00	0.00	53.216	0.000	0.00000
0.20	40.00	52.842	5.384	0.86437
0.40	80.00	50.553	18.476	1.37607
0.60	80.00	45.524	30.749	0.81461
0.80	80.00	38.584	38.014	0.48224
1.00	80.00	30.514	42.315	0.28548
1.20	80.00	21.774	44.862	0.16900
1.40	80.00	12.638	46.369	0.10005
1.60	80.00	3.267	47.261	0.05923
1.80	80.00	−6.242	47.789	0.03506
2.00	80.00	−15.834	48.102	0.02076
2.20	80.00	−25.475	48.287	0.01229
2.40	80.00	−35.144	48.397	0.00727
2.60	40.00	−44.457	43.078	−0.86007
2.80	0.00	−51.864	30.024	−1.37352

Maximum values using simplified method per Eqs. (3.5) and (3.6):

$$\dot{\phi} = 48.6 \text{ deg/s} \qquad \ddot{\phi} = 2.22 \text{ rad/s}^2$$
$$M_\delta \delta = 687,500 \text{ ft-lb} \qquad b/2V = 0.06361/s$$
$$M_\phi = 811,311.6 \text{ ft-lb} \qquad k_1 = 2.62136 \qquad k_2 = 0.02777$$

maximum aileron power control unit (PCU) capability; hence the time and wheel rates are selected accordingly.

3.6 Roll Termination Condition

The roll termination condition shown in Fig. 3.4, represented by the return to neutral of the lateral control wheel (time t_2 to t_3), is not required for analysis by current FAR 25 regulations.

For aircraft designed to other requirements, such as military usage, consideration must be given to the roll termination problem. In the examples given in Tables 3.5 and 3.6, the lateral controls are shown returned to neutral at the same rate as initially applied. For this example the acceleration is slightly less than the maximum obtained at t_1 (with the sign changed), but the roll velocity is still significant.

3.7 Nonlinear Lateral Control Inputs

An example of the lateral control wheel position vs surface motion is shown in Fig. 3.5. The spoiler input delay at 10 deg of wheel position is typical of configurations to eliminate or minimize lift loss due to spoilers during autopilot

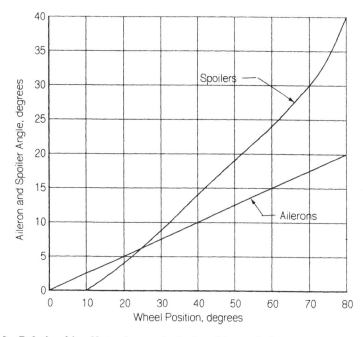

Fig. 3.5   Relationship of lateral control wheel position and aileron and spoiler angles.

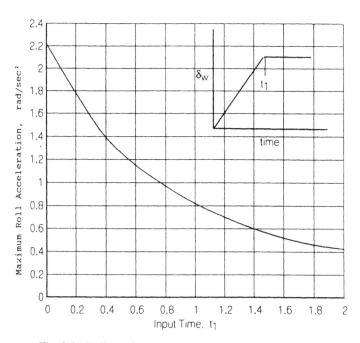

Fig. 3.6   Roll acceleration as a function of input time t_1.

Table 3.6 Rolling maneuver analysis using approximate solution by finite difference methods on a personal computer with linear lateral control surface motion defined in Fig. 3.4

$t_1, s = 0.4 \quad t_2, s = 2.4 \quad t_3, s = 2.8$

Time, s	Wheel, deg	Roll angle, deg	Rolling velocity, deg/s	Rolling acceleration, rad/s^2
0.00	0.00	53.216	0.000	0.00000
0.20	40.00	52.714	5.019	0.87583
0.40	80.00	50.417	17.951	1.38121
0.60	80.00	45.576	30.456	0.80103
0.80	80.00	38.755	37.755	0.47281
1.00	80.00	30.771	42.087	0.28328
1.20	80.00	22.094	44.682	0.16966
1.40	80.00	13.002	46.237	0.10158
1.60	80.00	3.662	47.167	0.06082
1.80	80.00	−5.827	47.724	0.03641
2.00	80.00	−15.405	48.058	0.02180
2.20	80.00	−25.037	48.258	0.01305
2.40	80.00	−34.700	48.377	0.00782
2.60	40.00	−43.881	43.430	−0.87115
2.80	−0.00	−51.278	30.540	−1.37841

Maximum values using simplified method per Eqs. (3.5) and (3.6):

$$\dot{\phi} = 48.6 \text{ deg/s} \qquad \ddot{\phi} = 2.22 \text{ rad/s}^2$$
$$M_\delta \delta = 687{,}500 \text{ ft-lb} \qquad b/2V = 0.06361/s$$
$$M_{\dot\phi} = 811{,}311.6 \text{ ft-lb} \qquad k_1 = 2.62136 \qquad k_2 = 0.02777$$

operation. These curves do not show any blowdown of the aileron and spoilers that may occur at high airspeeds.

This control system, although nonlinear with respect to wheel vs surface motion, may still be represented by assuming linear inputs.

As noted in Fig. 3.5, the aileron angle is linear with wheel input, but because of the delay in spoiler motion with wheel angle the resulting motion is nonlinear. Since design rolling maneuver conditions are for maximum wheel input, hence maximum aileron and spoiler angles (except as limited by blowdown limits), the assumption in assuming linear inputs is conservative.

3.8 Aeroelastic Effects

In modern jet transports with significant sweep in the wings, aeroelasticity has a pronounced effect on three of the parameters shown in the equation of motion, Eq. (3.3). These parameters are as follows:

The term $M_{\ddot\phi}$ is the rolling moment due to roll acceleration that is induced due to aeroelastic effects. This moment becomes zero for a rigid wing.

Table 3.7   Calculation of lateral control maximum input
rates as a function of t_1 for an airplane with a typical design
lateral control configuration[a]

t_1, s	δ_w,[b] deg	δ_w,[c] deg/s	δ_a,[d] deg	$\dot{\delta}_a$,[e] deg/s
0.3	80	267	20	67
0.4	80	200	20	50
0.5	80	160	20	40
0.6	80	133	20	33
1.0	80	80	20	20
1.5	80	53	20	13

[a]The time history analyses shown in Tables 3.5 and 3.6 are based on a maximum input rate of 50 deg/s for the ailerons such that the lateral control wheel reaches the maximum angle in 0.40 s.
[b]δ_w = maximum wheel angle available.
[c]δ_w = maximum wheel rate assuming linear input.
[d]δ_a = maximum aileron angle available.
[e]δ_a = maximum aileron rate assuming linear input.

The term $M_{\dot{\phi}}$ is the roll damping moment due to roll velocity that will induce an incremental rolling moment due to aeroelastic effects. This effect will reduce the damping moment for swept-back wings.

The term $M_\delta \delta$ is the rolling moment due to lateral control application that is reduced due to aeroelastic effects.

Per FAR 25.349(a) the effect of aeroelasticity must be considered in determining the required aileron deflections used for rolling maneuver analyses.

For swept-back wings with outboard ailerons it is possible to have a situation where the effect of aeroelasticity induced by pitching moment due to the aileron overcomes the effect of the lift increment producing airplane roll, thus creating the aeroelastic phenomenon called "aileron reversal." This is discussed in Chapter 12.

References

[1]Etkin, B., *Dynamics of Flight, Stability and Control,* Wiley, New York, 1959.

[2]Hildebrand, F. B., *Advanced Calculus for Engineers,* Prentice–Hall, Englewood Cliffs, NJ, 1949.

[3]Wylie, C. R., Jr., *Advanced Engineering Mathematics,* McGraw–Hill, New York, 1960.

4
Yawing Maneuvers

Yawing maneuvers when applied to structural load analyses are maneuvers involving the abrupt application of the rudder in producing a sideslip condition or during engine-out conditions.

Two types of yawing maneuvers, as shown schematically in Fig. 4.1, must be considered for structural design.

1) Rudder maneuvers as used for structural design are essentially flat maneuvers whereby the rudder is abruptly applied in a wings-level attitude. This maneuver is difficult to do in flight because large amounts of lateral control must be applied to maintain wings level. The purpose of holding the wings level is to maximize the resulting sideslip.

2) Engine-out maneuvers, as used for structural design, are essentially flat maneuvers whereby abrupt application of the rudder is made in conjunction with the resulting sideslip due to unsymmetrical engine thrust.

4.1 Parameters Required for Structural Load Analyses

The parameters required for structural load analyses for yawing maneuver conditions are shown in Table 4.1. The relationships between applied rudder and sideslip for rudder maneuvers, and between unsymmetrical thrust, sideslip, and corrective rudder for engine-out maneuvers, may be determined using the methods developed in this chapter.

Asymmetrical conditions are defined as the incremental loads due to roll or yaw before the inclusion of the symmetrical flight load increments.

4.2 Rudder Maneuver Requirements—FAR 25 Criteria

The rudder maneuver requirements of FAR 25.351(a)[1] are as follows.

At speeds from V_{MC} to V_D, the following maneuvers must be considered. In computing the tail loads, the yawing velocity may be assumed to be zero.

1) Maneuver I: With the airplane in unaccelerated flight at zero yaw, it is assumed that the rudder control is suddenly displaced to the maximum deflection as limited by the control stops or by a 300-lb rudder pedal force, whichever is less.

2) Maneuver II: With the rudder deflected as specified in FAR 25.351(a)(1), it is assumed that the airplane yaws to the resulting sideslip.

3) Maneuver III: With the airplane yawed to the static sideslip angle corresponding to the rudder deflection specified in FAR 25.351(a)(1), it is assumed that the rudder is returned to neutral.

Those maneuvers, shown schematically in Fig. 4.2, are labeled for convenience of analysis as maneuvers I, II, and III.

The only difference between JAR 25.351(a) and FAR 25.351(a) is that the pilot force applied to the rudder pedals between V_C/M_C and V_D/M_D is reduced to 200 lb. On modern jet transports with control systems, this becomes academic because of the ability to obtain full available rudder with less than 200 lb. The rudder is

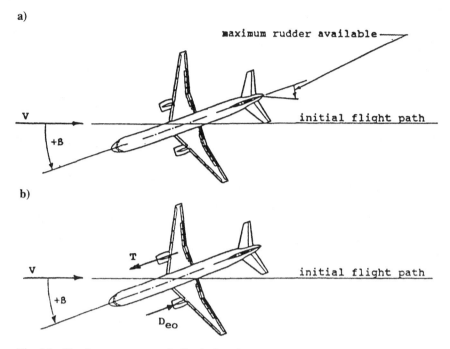

Fig. 4.1   Yawing maneuvers: a) pilot-induced rudder maneuvers and b) asymmetrical thrust (engine-out) maneuvers.

limited only by the amount of hinge moment the power control unit (PCU) can produce or by the rudder stops.

4.2.1   Rudder Maneuver Requirements for the Military Criteria

It should be noted that there is a significant difference in philosophy between the rudder maneuver criteria of FAR and JAR[2] and U.S. military specifications (MIL-A-8861)[3] for transport-type aircraft. The basic maneuver conditions are the same, but the definition of the available rudder is different for the overyaw and steady sideslip conditions.

The commercial regulations stipulate that the amount of sideslip obtained is determined from the amount of rudder deflection obtained in the maneuver I condition. For military specifications the amount of sideslip used for maneuvers II and III conditions is based on the amount of pedal force applied for the maneuver I condition. This difference is subtle, in that the rudder for maneuvers II and III may become greater with sideslip, depending on the design characteristics of the directional control system.

Table 4.1   Unsymmetrical load analysis parameters required

Unsymmetrical condition	Symmetrical condition load factor	Asymmetrical condition
Yaw maneuvers	1.0	Yaw rates and accelerations Rudder and sideslip angles

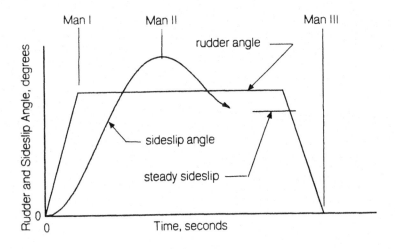

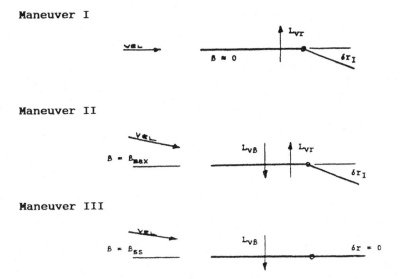

Fig. 4.2 Definition of rudder maneuvers; load vectors are shown in the absolute sense.

This question can be raised: "Are the commercial regulations adequate from a rudder maneuver standpoint?" The answer is that the use of the commercial regulations has proven adequate for structural design over the years and that the criteria as applied are based on simplistic maneuvers that are difficult to achieve in actual service operation of commercial aircraft.

4.2.2 Rudder Available

The magnitude of the maximum rudder deflection available at any given airspeed and Mach number will be a function of the pilot effort plus the PCU that can be applied directly to the rudder. Within the criteria stated in Sec. 4.2, the maximum

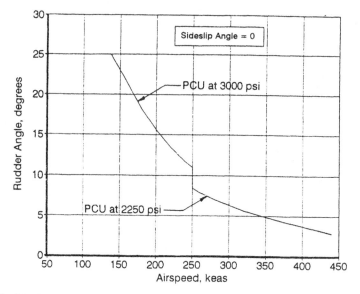

Fig. 4.3   Maximum rudder available with a two-stage system pressure reducer limiting device activated at 250 kcas.

available rudder obtainable as a function of airspeed must be used for design. The method for determining design control surface angles is discussed in Sec. 11.3 in Chapter 11.

4.2.3   Rudder Limiting Devices

The amount of rudder available in older aircraft designed per Civil Air Regulations (CAR 4b)[4] was limited by the pilot effort requirements for the type of control system considered in the design. With the advent of fully powered surfaces with multiple power control units, means of limiting the amount of rudder available at high speeds must be considered to reduce the structural weight of the vertical tails and aft body structure.

Examples of two system designs that limit the amount of rudder available at high speeds are discussed herein.

1) The first is a power control unit having full system pressure below a given speed and a reduced pressure across the PCU pistons for high-speed conditions. The rudder available for this type of system is shown in Fig. 4.3.

2) The second is a power control unit limited such that the amount of rudder available is restricted by a ratio changer that is a function of calibrated airspeed. The rudder available for this type of system is shown in Fig. 4.4. The advantage of this type of system is that the rudder can be programmed as a function of airspeed such that the aerodynamic requirements are met for engine-out and crosswind landings and the impact on structural loads is minimized.

4.2.4   Steady Sideslip Due to Rudder

Before the flight test of a new airplane, steady sideslips due to rudder may be calculated from wind-tunnel data corrected for full-scale airplane effects such as aeroelasticity. Assuming that the airplane is in level flight at a constant airspeed

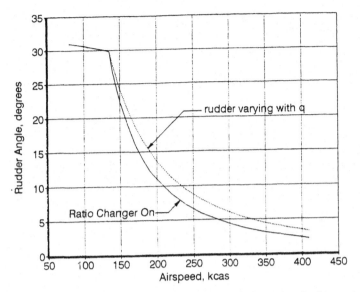

Fig. 4.4 **Maximum rudder available as limited by a ratio changer in the control system; the pilot can obtain maximum available rudder at the maximum pedal movement regardless of airspeed. The dotted curve shows the rudder available as a function of q if the limiting device were not incorporated into the rudder system.**

and the aerodynamic coefficients are linear, the three equations for side force, yawing, and rolling moments may be written in coefficient form, using matrix notation:

$$
\begin{bmatrix} Cy_\beta & Cy_{\delta w} & -C_L \\ Cn_\beta & Cn_{\delta w} & 0 \\ Cl_\beta & Cl_{\delta w} & 0 \end{bmatrix} \begin{bmatrix} \beta_{ss} \\ \delta_w \\ \phi \end{bmatrix} = - \begin{bmatrix} Cy_{\delta r} \\ Cn_{\delta r} \\ Cl_{\delta r} \end{bmatrix} \delta r \tag{4.1}
$$

where β, δr, and δw are the sideslip, rudder, and control wheel angles, respectively; Cy_β, $Cy_{\delta r}$, and $Cy_{\delta w}$ are the side force coefficient due to sideslip, rudder, and aileron/spoilers, respectively; Cn_β, $Cn_{\delta r}$, and $Cn_{\delta w}$ are the yawing moment coefficient due to sideslip, rudder, and aileron/spoilers, respectively; Cl_β, $Cl_{\delta r}$, and $Cl_{\delta w}$ are the rolling moment coefficient due to sideslip, rudder, and aileron/spoilers, respectively; C_L is the airplane lift coefficient; and ϕ is the airplane bank angle.

For aircraft with spoilers used for lateral control in conjunction with ailerons, the relationship with wheel angle may be written as shown in Eq. (4.2) assuming linear derivatives:

$$
Cn_{\delta w} \delta w = Cn_{\delta a} \delta_a + Cn_{\delta sp} \delta_{sp} \tag{4.2}
$$

As is the case for many aircraft, the variation of spoiler angle with aileron angle is not linear due to the delay in spoilers with wheel angle (usually due to autopilot considerations). This is shown in Fig. 3.5. Traditionally the relationship may be assumed linear for solution of the rudder maneuver problem.

Since the rudder available for the condition under consideration is known, solution of Eq. (4.1) may be accomplished to obtain sideslip, wheel, and bank angles for steady sideslips:

$$\beta_{ss} = \frac{(-Cn_{\delta r} + Cl_{\delta r} Cn_{\delta w}/Cl_{\delta w})}{Cn_\beta - Cl_\beta Cn_{\delta w}/Cl_{\delta w}} \delta r \qquad (4.3)$$

$$\delta_w = (-Cl_{\delta r}\delta r - Cl_\beta \beta_{ss})/Cl_{\delta w} \qquad (4.4)$$

$$\phi = (Cy_\beta \beta_{ss} + Cy_{\delta r}\delta r + Cy_{\delta w}\delta w)/C_L \qquad (4.5)$$

4.3 Engine-Out Maneuver Requirements—FAR 25 Criteria

The engine-out maneuver requirements of FAR 25.367 are as follows.

1) The airplane must be designed for the unsymmetrical loads resulting from failure of the critical engine. Turbopropeller airplanes must be designed for the following conditions in combination with a single malfunction of the propeller drag limiting system, considering the probable pilot corrective action on the flight controls.

a) At speeds between V_{MC} and V_D, the loads resulting from power failure because of fuel flow interruption are considered to be limit loads.

b) At speeds between V_{MC} and V_C, the loads resulting from the disconnection of the engine compressor from the turbine or from loss of the turbine blades are considered to be ultimate loads.

c) The time history of the thrust decay and drag buildup occurring as a result of the prescribed engine failures must be substantiated by test or other data applicable to the particular engine–propeller combination.

d) The timing and magnitude of the probable pilot corrective action must be conservatively estimated, considering the characteristics of the particular engine–propeller–airplane combination.

2) Pilot corrective action may be assumed to be initiated at the time maximum yawing velocity is reached, but not earlier than 2 s after the engine failure. The magnitude of the corrective action may be based on the control forces specified in FAR 25.397(b) except that lower forces may be assumed where analysis or test shows that these forces can control the yaw and roll resulting from the prescribed engine failure conditions.

It should be noted that there is no significant difference between FAR 25.367 and JAR 25.367.

In essence, two conditions are defined in the preceding regulations at each of the airspeeds under consideration: 1) maximum sideslip produced with zero rudder and 2) corrective rudder applied no sooner than 2 s or at time of maximum yaw rate. The amount of corrective rudder is not specified, but usually the rudder required for zero sideslip in the engine-out steady-state condition is used.

Two basic design conditions must be considered.

1) Engine-out conditions due to fuel flow interruption are to be considered a limit load condition, and hence a safety factor of 1.5 must be applied.

2) Engine-out conditions due to mechanical failure of the engine or propeller systems are to be considered as an ultimate load condition, and hence a safety factor of 1.0 must be applied.

The difference between these two conditions is the time of engine thrust decay. Fuel flow interruption may occur from 1 s to as long as several seconds, whereas a mechanical failure happens very abruptly. Examples of thrust decay are shown in Fig. 4.5.

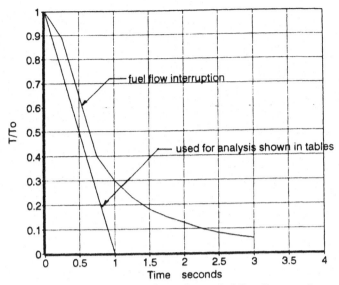

Fig. 4.5   Typical thrust decay due to fuel flow interruption.

4.3.1   Engine-Out Analysis Steady-State Conditions

Two conditions must be considered in solving the steady-state engine-out problem: 1) the maximum steady sideslip with zero rudder and 2) the rudder required to balance the engine-out at zero sideslip.

The amount of yawing moment due to engine-out may be determined from Eq. (4.6), considering the thrust on the remaining engines and the drag of the dead engine:

$$Cn_{eo} = (T + D_{eo})a_{eo}/q S_w b_w \tag{4.6}$$

where T is the engine net thrust (lb), D_{eo} is the drag of the dead engine (lb), and a_{eo} is the arm of the dead engine (in.).

Equation (4.6) is written assuming a single failure in which the thrust and dead engine drag are acting on the appropriate opposite engines; hence the equation as shown is applicable to a configuration with more than two engines.

The steady-state equations summarized in Eq. (4.7) for engine-out conditions are determined in a similar manner to the rudder maneuver analysis shown in Sec. 4.2.4:

$$\begin{bmatrix} Cy_\beta & Cy_{\delta w} & Cy_{\delta r} \\ Cn_\beta & Cn_{\delta w} & Cn_{\delta r} \\ Cl_\beta & Cl_{\delta w} & Cl_{\delta r} \end{bmatrix} \begin{bmatrix} \beta_{eo} \\ \delta_w \\ \delta_r \end{bmatrix} = \begin{bmatrix} C_L\phi \\ -Cn_{eo} \\ 0 \end{bmatrix} \tag{4.7}$$

4.3.2   Engine-Out Steady Sideslip with Zero Rudder

If the assumption is made that the rudder is held neutral, then the steady sideslip parameters for an engine-out condition may be determined from the solution of Eq. (4.7):

$$\beta_{eo} = \frac{-Cn_{eo}}{Cn_\beta - Cl_\beta Cn_{\delta w}/Cl_{\delta w}} \tag{4.8}$$

$$\delta_w = -Cl_\beta \beta_{eo}/Cl_{\delta w} \tag{4.9}$$

$$\phi_{eo} = (Cy_\beta \beta_{eo} + Cy_{\delta w}\delta_w)/C_L \tag{4.10}$$

4.3.3 Engine-Out Steady Condition with Zero Sideslip

If the assumption is made that sideslip is zero, then the rudder to balance an engine-out condition may be determined from the solution of Eq. (4.7):

$$\delta_{reo} = \frac{-Cn_{eo}}{Cn_{\delta r} - Cl_\beta Cn_{\delta w}/Cl_{\delta w}} \tag{4.11}$$

$$\delta_w = -Cl_{\delta reo}\delta r/Cl_{\delta w} \tag{4.12}$$

$$\phi_{eo} = (Cy_{\delta r}\delta_{reo} + Cy_{\delta w}\delta_w)/C_L \tag{4.13}$$

4.4 Equations of Motion for Yawing Maneuvers

A complete set of the equations of motion for an airplane involving side translation, yaw, and roll degrees of freedom is derived in Ref. 5.

Since the yawing maneuver used for structural load analyses is considered to be a "flat maneuver," the following assumptions are made:

1) Roll acceleration and velocity are assumed zero.

2) Lateral control is applied as necessary to maintain a wings-level attitude.

3) Airspeed and Mach number (hence altitude) are assumed constant during the maneuver.

4) Rate derivatives of the rudder and lateral control devices are neglected.

By dividing the side force equation by qS_w and the roll and yaw moment equations by qS_wb_w, one can derive the equations of motion for yawing maneuvers. The equation written in matrix notation in Eq. (4.14) is called the three degree-of-freedom (DOF) method in this book:

$$\begin{bmatrix} a_1 & -Cn_{\delta w} & 0 \\ a_2 & -Cl_{\delta w} & 0 \\ 0 & -Cy_{\delta w} & a_3 \end{bmatrix} \begin{bmatrix} \ddot{\psi} \\ \delta_w \\ \dot{\beta} \end{bmatrix} = \begin{bmatrix} b_1 \\ b_2 \\ b_3 \end{bmatrix} \tag{4.14}$$

where

$$a_1 = I_z/(qS_wb_w) \tag{4.15}$$

$$a_2 = -I_{xz}/(qS_wb_w) \tag{4.16}$$

$$a_3 = MV_T/(qS_w) \tag{4.17}$$

$$b_1 = Cn_{eo} + Cn_{\delta r}\delta_r + Cn_r\dot{\psi} + Cn_\beta\beta \tag{4.18}$$

$$b_2 = Cl_{\delta r}\delta_r + Cl_r\dot{\psi} + Cl_\beta\beta \tag{4.19}$$

$$b_3 = Cy_{\delta r}\delta_r - [MV_T/(qS_w) - Cy_r]\dot{\psi} + Cy_\beta\beta \tag{4.20}$$

To allow for the possible use of spoilers along with ailerons for lateral control (which is normally the case for the large commercial jets in operation today), Eq. (4.14) is written in terms of wheel angle using the relationships shown in Eq. (4.2).

A further simplification of Eq. (4.14) is shown in Table 4.2 whereby the equations of motion are reduced to two DOF, neglecting the roll degree of freedom. The differences in the results for the two methods of analysis will be discussed in Sec. 4.4.1.

Table 4.2   Yawing maneuver analysis—equations of motion, Eq. (4.14),
modified assuming two-DOF method whereby the roll degree
of freedom is neglected

$$\begin{bmatrix} a_1 & 0 & 0 \\ 0 & -Cy_{\delta w} & a_3 \end{bmatrix} \begin{bmatrix} \ddot{\psi} \\ \dot{\beta} \end{bmatrix} = \begin{bmatrix} b_1 \\ b_3 \end{bmatrix} \qquad (4.14a)$$

where

$$a_1 = I_z/(qS_w b_w) \qquad (4.14b)$$
$$a_3 = MV_T/(qS_w) \qquad (4.14c)$$
$$b_1 = Cn_{e_0} + Cn_{\delta r}\delta_r + Cn_r\dot{\psi} + Cn_\beta\beta \qquad (4.14d)$$
$$b_3 = Cy_{\delta r}\delta_r - [MV_T/(qS_w) - Cy_r]\dot{\psi} + Cy_\beta\beta \qquad (4.14e)$$

4.4.1   Rudder Maneuver Analyses

Solution of Eq. (4.14) for a rudder maneuver (neglecting the engine-out term) will give the response of the airplane to the rudder inputs defined in Sec. 4.2.

Assuming linear aerodynamic coefficients and using numerical integration techniques,[6,7] Tables 4.3 and 4.4 show rudder maneuver analyses for an airplane using the two methods of analysis represented by Eqs. (4.14) and (4.14a). These analyses show the abrupt rudder condition, maneuver I, and the maximum overyaw condition, maneuver II.

The airplane response for the maneuver III condition, which is the abrupt rudder return to neutral from the steady sideslip condition, may be determined using the maneuver I response superimposed onto the steady sideslip condition.

The differences between the three-DOF and two-DOF analyses are as follows: the maneuver I condition reveals an insignificant difference in sideslip between the two analyses, and in the maneuver II condition, the overyaw angle increases by 0.8% for the simplified method.

Although the time to set up and run the two methods on a personal computer is not significantly different, the simplified method requires less data input for the analysis.

4.4.2   Engine-Out Maneuver Analyses

Solution of Eq. (4.14) for an engine-out maneuver will give the response of the airplane to the rudder inputs defined in Sec. 4.3. The engine-out yawing moment coefficient is defined by Eq. (4.6).

It usually is necessary to solve the engine-out analysis in two steps, as follows.

1) Using a thrust decay as the input, and assuming no corrective action by the pilot with rudder, one runs the time history analysis to determine the maximum sideslip angle of the airplane and the time of maximum yaw rate.

2) The analysis is now rerun with corrective rudder initiated at the time of maximum yaw rate or no sooner than 2 s. The corrective rudder shown in Table 4.5 is as required to produce zero sideslip in the steady-state condition. It should be noted that for the example shown the aircraft reached maximum yaw rate before the 2.0-s time stipulated as the minimum time for corrective action by the pilot.

Assuming linear aerodynamic coefficients and using numerical integration techniques,[6,7] Table 4.5 shows an engine-out analysis for the condition without corrective rudder action. Table 4.6 shows an engine-out analysis whereby corrective rudder action is initiated at 2.0 s. The amount of rudder applied in this example is as required to produce zero sideslip in the steady-state condition.

Table 4.3   Rudder maneuver analysis based on the solution of Eq. (4.14) assuming three DOF and linear rudder input motion

Cond.: 1
  Alt., ft $= 0$ V_e, keas $= 240$ Mach $= 0.363$
  Gross weight, lb $= 219{,}000$ CG, % mac/100 $= 0.352$

Steady sideslip solution

δ_w/δ_r	β/δ_r	β, deg	δ_w, deg	ϕ, deg	$\delta_{r\,max}$, deg
3.782	0.815	5.869	27.231	−4.488	7.200

Time, s	δ_r, deg	δ_w, deg	β, deg	$\ddot{\psi}$, rad/s²	$\dot{\psi}$, rad/s	$\dot{\beta}$, rad/s
Time history analysis ($\Delta t = 0.02$ s)						
0.20	7.20	−3.05	0.102	−0.124	−0.015	0.019
0.40	7.20	−0.59	0.469	−0.110	−0.038	0.042
0.60	7.20	3.05	1.071	−0.092	−0.059	0.061
0.80	7.20	7.59	1.862	−0.071	−0.075	0.075
1.00	7.20	12.77	2.789	−0.048	−0.086	0.085
1.20	7.20	18.28	3.799	−0.024	−0.093	0.090
1.40	7.20	23.82	4.834	−0.000	−0.095	0.090
1.60	7.20	29.13	5.844	0.021	−0.093	0.086
1.80	7.20	33.96	6.781	0.040	−0.087	0.078
2.00	7.20	38.12	7.604	0.056	−0.077	0.066
2.20	7.20	41.43	8.281	0.068	−0.064	0.053
2.40	7.20	43.81	8.789	0.076	−0.049	0.037
2.60	7.20	45.20	9.115	0.080	−0.034	0.021
2.80	7.20	45.60	9.257	0.079	−0.018	0.005
3.00	7.20	45.07	9.220	0.075	−0.003	−0.010
3.20	7.20	43.69	9.019	0.067	0.012	−0.024
3.40	7.20	41.60	8.675	0.056	0.024	−0.035
3.60	7.20	38.94	8.214	0.044	0.034	−0.044
3.80	7.20	35.89	7.669	0.030	0.041	−0.050
4.00	7.20	32.62	7.071	0.016	0.045	−0.053
Time of maximum rudder—maneuver I						
0.18	7.20	−3.22	0.079	−0.125	−0.013	0.017
Time of maximum overyaw—maneuver II						
2.86	7.20	45.54	9.264	0.078	−0.013	0.001

Table 4.4   Rudder maneuver analysis based on the solution of Eq. (4.14) in Table 4.2 assuming two DOF and linear rudder input motion

Cond.: 2
  Alt., ft = 0 V_e, keas = 240   Mach = 0.363
  Gross weight, lb = 219,000 CG, % mac/100 = 0.352

			Steady sideslip solution			
δ_w/δ_r		β/δ_r	β, deg	δ_w, deg	ϕ, deg	$\delta_{r\,max}$, deg
0.000		0.825	5.941	0.000	0.000	7.200

Time, s	δ_r, deg	δ_w, deg	β, deg	$\ddot{\psi}$, rad/s²	$\dot{\psi}$, rad/s	$\dot{\beta}$, rad/s
			Time history analysis ($\Delta t = 0.02$ s)			
0.20	7.20	0.00	0.103	−0.123	−0.015	0.020
0.40	7.20	0.00	0.473	−0.110	−0.038	0.042
0.60	7.20	0.00	1.080	−0.092	−0.058	0.061
0.80	7.20	0.00	1.877	−0.071	−0.075	0.076
1.00	7.20	0.00	2.811	−0.048	−0.086	0.086
1.20	7.20	0.00	3.827	−0.024	−0.093	0.091
1.40	7.20	0.00	4.870	−0.001	−0.095	0.091
1.60	7.20	0.00	5.887	0.021	−0.093	0.086
1.80	7.20	0.00	6.830	0.040	−0.087	0.078
2.00	7.20	0.00	7.660	0.055	−0.077	0.067
2.20	7.20	0.00	8.342	0.067	−0.065	0.053
2.40	7.20	0.00	8.856	0.075	−0.050	0.038
2.60	7.20	0.00	9.187	0.079	−0.035	0.022
2.80	7.20	0.00	9.333	0.078	−0.019	0.006
3.00	7.20	0.00	9.300	0.074	−0.004	−0.010
3.20	7.20	0.00	9.103	0.066	0.010	−0.023
3.40	7.20	0.00	8.762	0.056	0.022	−0.035
3.60	7.20	0.00	8.304	0.044	0.032	−0.044
3.80	7.20	0.00	7.761	0.030	0.040	−0.050
4.00	7.20	0.00	7.164	0.016	0.044	−0.053
			Time of maximum rudder—maneuver I			
0.18	7.20	0.00	0.080	−0.125	−0.013	0.017
			Time of maximum overyaw—maneuver II			
2.86	7.20	0.00	9.342	0.077	−0.014	0.001

**Table 4.5   Engine-out maneuver analysis based on the solution of
Eq. (4.14) assuming three DOF and no corrective rudder action**

Cond.: 3

Alt., ft $= 0$ V_e, keas $=240$ Mach $= 0.363$

Gross weight, lb $= 219,000$ CG, % mac/100 $= 0.352$

T, lb $= 24,800$ D_{eo}, lb $= -1200$

$Cn_{eo\,max} = 0.01188$ Thrust decay, s $= 1.00$

Steady sideslip solution with rudder $= 0$:

β, deg $= -3.414$ δ_w, deg $= -18.49$ ϕ, deg $= 4.59$

Rudder required to balance engine-out at $\beta = 0$:

δ_r, deg $= 4.188$ δ_w, deg $= -2.65$ ϕ, deg $= 1.98$

Time, s	δ_r, deg	δ_w, deg	β, deg	$\ddot{\psi}$, rad/s^2	$\dot{\psi}$, rad/s	Cn_{eo}
		Time history analysis $(\Delta t = 0.02\,\text{s})$				
0.20	0.00	-0.15	-0.007	0.015	0.002	0.00238
0.40	0.00	-0.55	-0.050	0.028	0.006	0.00475
0.60	0.00	-1.33	-0.156	0.039	0.013	0.00713
0.80	0.00	-2.62	-0.348	0.047	0.022	0.00950
1.00	0.00	-4.49	-0.645	0.054	0.032	0.01188
1.20	0.00	-6.84	-1.048	0.042	0.041	0.01188
1.40	0.00	-9.58	-1.535	0.030	0.048	0.01188
1.60	0.00	-12.55	-2.076	0.017	0.053	0.01188
1.80	0.00	-15.59	-2.642	0.004	0.055	0.01188
2.00	0.00	-18.55	-3.201	-0.008	0.054	0.01188
2.20	0.00	-21.29	-3.728	-0.019	0.052	0.01188
2.40	0.00	-23.68	-4.200	-0.028	0.047	0.01188
2.60	0.00	-25.65	-4.596	-0.036	0.040	0.01188
2.80	0.00	-27.11	-4.904	-0.041	0.033	0.01188
3.00	0.00	-28.04	-5.113	-0.043	0.024	0.01188
3.20	0.00	-28.43	-5.222	-0.043	0.015	0.01188
3.40	0.00	-28.29	-5.231	-0.042	0.007	0.01188
3.60	0.00	-27.67	-5.148	-0.038	-0.001	0.01188
3.80	0.00	-26.64	-4.983	-0.033	-0.008	0.01188
4.00	0.00	-25.27	-4.749	-0.026	-0.014	0.01188
		Time of maximum overyaw—maneuver II				
3.32	0.00	-28.40	-5.239	-0.043	0.010	0.01188
		Time of maximum yaw rate				
1.86	0.00	-16.50	-2.811	0.000	0.055	0.01188

Table 4.6 Engine-out maneuver analysis based on the solution of Eq. (4.14) assuming three DOF and corrective rudder action initiated at 2.0 s; the corrective rudder is as required for zero sideslip in the steady-state condition

Cond.: 4

Alt., ft $= 0$ V_e, keas $= 240$ Mach $= 0.363$

Gross weight, lb $= 219{,}000$ CG, % mac/100 $= 0.352$

T, lb $= 24{,}800$ D_{eo}, lb $= -1200$

$Cn_{eo\,max} = 0.01188$ Thrust decay, s $= 1.00$

Steady sideslip solution with rudder $= 0$:

β, deg $= -3.414$ δ_w, deg $= -18.49$ ϕ, deg $= 4.59$

Rudder required to balance engine-out at $\beta = 0$:

δ_r, deg $= 4.188$ δ_w, deg $= -2.65$ ϕ, deg $= 1.98$

Time, s	δ_r, deg	δ_w, deg	β, deg	$\ddot{\psi}$, rad/s²	$\dot{\psi}$, rad/s	Cn_{eo},
		Time history analysis ($\Delta t = 0.02$ s)				
0.20	0.00	−0.15	−0.007	0.015	0.002	0.00238
0.40	0.00	−0.55	−0.050	0.028	0.006	0.00475
0.60	0.00	−1.33	−0.156	0.039	0.013	0.00713
0.80	0.00	−2.62	−0.348	0.047	0.022	0.00950
1.00	0.00	−4.49	−0.645	0.054	0.032	0.01188
1.20	0.00	−6.84	−1.048	0.042	0.041	0.01188
1.40	0.00	−9.58	−1.535	0.030	0.048	0.01188
1.60	0.00	−12.55	−2.076	0.017	0.053	0.01188
1.80	0.00	−15.59	−2.642	0.004	0.055	0.01188
2.00	0.00	−18.55	−3.201	−0.008	0.054	0.01188
2.20	4.20	−22.88	−3.644	−0.090	0.040	0.01188
2.40	4.20	−23.70	−3.873	−0.091	0.022	0.01188
2.60	4.20	−23.42	−3.895	−0.087	0.004	0.01188
2.80	4.20	−22.15	−3.724	−0.079	−0.013	0.01188
3.00	4.20	−20.01	−3.380	−0.068	−0.027	0.01188
3.20	4.20	−17.16	−2.894	−0.055	−0.039	0.01188
3.40	4.20	−13.80	−2.299	−0.039	−0.049	0.01188
3.60	4.20	−10.12	−1.632	−0.023	−0.055	0.01188
3.80	4.20	−6.34	−0.931	−0.007	−0.058	0.01188
4.00	4.20	−2.63	−0.231	0.008	−0.057	0.01188
		Time of maximum overyaw				
2.52	4.20	−23.66	−3.911	−0.089	0.011	0.01188
		Time of maximum yaw rate				
1.86	0.00	−16.50	−2.811	0.000	0.055	0.01188

References

[1]Anon., "Part 25—Airworthiness Standards: Transport Category Airplanes," Federal Aviation Regulations, U.S. Dept. of Transportation, Jan. 1994.

[2]Anon., "JAR 25 Large Aeroplanes," Joint Aviation Requirements, Oct. 1989.

[3]Anon., "Airplane Strength and Rigidity—Flight Loads," Military Specification, MIL-A-8861(ASG), May 1960.

[4]Anon., "Part 4b—Airplane Airworthiness Transport Categories," Civil Air Regulations, Civil Aeronautics Board, Dec. 1953.

[5]Etkin, B., *Dynamics of Flight, Stability and Control,* Wiley, New York, 1959.

[6]Hildebrand, F. B., *Advance Calculus for Engineers,* Prentice–Hall, Englewood Cliffs, NJ, 1949.

[7]Wylie, C. R., Jr., *Advance Engineering Mathematics,* McGraw–Hill, New York, 1960.

Flight in Turbulence

The gust criteria set forth by the certifying agencies for design of commercial aircraft have evolved over the years since the Douglas DC-6, Lockheed Constellation, and Boeing 377 Stratocruiser were designed. For further background on flight loads in turbulence the author recommends Ref. 1 as an exposition of the current engineering practice for calculating gust loads on aircraft.

5.1 Sharp-Edge Gust Criteria Based on Wing Loading

The earliest encounter by this author of gust analysis criteria was during the design phase of the early 707-100 airplane. With the proposal by the certifying agencies of revised gust criteria for the certification of the 707, review was made of the existing regulations and the impact the proposed new criteria would have on the DC-6 and 377 aircraft, if they were designed to it.

Before March 1956 the gust criteria in CAR 4b[2] (see Table 5.1) were based on the sharp-edge gust load factor, as shown in Eq. (5.1), with the gust factor K being defined as a function of wing loading W/S:

$$n_z = 1 + KUVC_{N\alpha}S_w/(575W) \qquad (5.1)$$

where V is the airplane airspeed (mph), $C_{N\alpha}$ is the slope of the airplane normal force coefficient curve vs wing angle of attack (per rad), and S is the wing reference area (ft^2).

The gust factor K is defined by either Eq. (5.2) or Eq. (5.3), depending on wing loading:

For $W/S_w < 16$ lb/ft^2:

$$K = 0.5(W/S_w)^{0..25} \qquad (5.2)$$

For $W/S > 16$ lb/ft^2:

$$K = 1.33 - 2.67/(W/S_w)^{0.75} \qquad (5.3)$$

It can be determined by inspection of Eq. (5.3) that the gust factor exceeds 1.0 at wing loadings greater than 16 lb/ft^2 and is less than 1.0 at wing loadings less than 16 lb/ft^2.

The gust factor K was determined from empirical calculations based on load factors measured on six aircraft[3] with sea level mass ratios from 6 to 23 and wing loadings from 5.4 to 44.5 lb/ft^2. The gust factors were normalized to $K = 1$ for the Boeing B-247 airplane, which had a wing loading of 16 lb/ft^2. The resulting curves defined by Eqs. (5.2) and (5.3) were considered applicable to all other aircraft that were similar to the B-247 airplane but had different wing loadings.[3]

Table 5.1   Gust velocities per CAR 4b
before March 1956

Design airspeeds	Gust Velocity U, ft/s
V_B	40
V_C	30
V_D	15

5.2 Revised Gust Criteria Using Airplane Mass Ratio

As aircraft size, wing shapes, and airspeeds increased, research, summarized in Refs. 4 and 5, indicated that gust loads were more closely a function of airplane mass parameter than of wing loading. The "one-minus-cosine 25 chord" discrete gust shape was also introduced in those reports.

Effective March 13, 1956, the criteria in CAR 4b were changed by Amendment 4b-3 to reflect this research. The revised discrete gust criteria were used on the 707-100/200/300, 727-100/200, and 737-100/200 aircraft. Other commercial transports of this time period, such as the DC-8 and Convair 880 aircraft, were also designed to the new criteria.

The gust load factors are determined from Eq. (5.4), which has become known as the "gust formula" method of analysis. This equation is similar to Eq. (5.1), except KU is replaced with $K_g U_{de}$:

$$n_z = 1 + K_g U_{de} V_e C_{N\alpha} S_w / (498W) \qquad (5.4)$$

where V_e is the airplane airspeed (keas), $C_{N\alpha}$ and S are as previously defined in this chapter, and K_g is the gust alleviation factor,

$$K_g = 0.88\mu_g / (5.3 + \mu_g) \qquad (5.5)$$

where μ_g is the airplane mass ratio in pitch,

$$\mu_g = 2W / (\rho c S_w g C_{N\alpha}) \qquad (5.6)$$

and where ρ is the density of air (slug/ft³), and c is the mean geometric chord of the wing (ft).

The gust alleviation factor K_g, defined in Eq. (5.5), was determined as an empirical function of airplane mass ratio to match the numerical results of the analysis shown in Ref. 5.

5.2.1 Design Gust Velocities

The design gust velocities based on the revised criteria using mass ratio are summarized in Table 5.2 and Fig. 5.1. As a matter of historical record the release of FAR 25 in December 1964, which was essentially a recodified CAR 4b, did not change the revised gust criteria per Amendment 4b-3.

5.2.2 Discrete Gust Shape

With the inclusion of the new criteria in CAR 4b and Amendment 4b-3 (and FAR 25), the shape of the gust to be used for analytical purposes was introduced. This shape, known as the one-minus-cosine 25 chord gust, became the basis

**Table 5.2 Gust velocities per CAR 4b
Amendment 4b-3, 1956**

Design Airspeeds	Gust velocity U_{de},[a] ft/s eas	
	Sea level to 20,000 ft	50,000 ft
V_B	66	38
V_C	50	25
V_D	25	12.5

[a]Linear variation between altitudes (see Fig. 5.1).

for discrete gust dynamic analysis until the introduction of power spectral density (PSD) requirements. Using the definition given in FAR 25.341(b), one can describe the shape of the gust in Eq. (5.7):

$$U(s) = (U_{de}/2)\left[1 - \cos\left(\frac{2\pi s}{25c}\right)\right] \qquad (5.7)$$

where s is the distance penetrated into the gust (ft), c is the mean geometric chord of the wing (ft), and U_{de} is the derived equivalent gust velocity (ft/s eas).

5.2.3 Comparison Between Old and Revised Criteria

A comparison between the old and revised criteria of CAR 4b is shown in Table 5.3 for the Boeing 377 Stratocruiser.

It is interesting to note that although the gust velocities increased from the old criteria to the revised, the product of $\bar{x}U$ and $K_g U_{de}$ that affected the resulting load factor is not significantly changed. This was intentional because the original

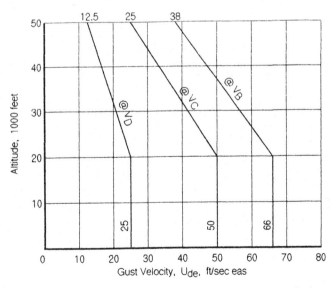

Fig. 5.1 Gust velocities for revised discrete gust criteria using airplane mass ratio [FAR 25.341(a)].

Table 5.3 Comparison of gust design criteria
before and after Amendment 4b-3 to CAR 4b
using the 377 Stratocruiser as an example[a]

S_w, ft^2 = 1710	mac, ft = 12.87	
Alt., ft	0	25,000
V_e, mph eas	312	300
Mach	0.41	0.65
$C_{L\alpha}$, per rad	5.186	6.824
W, lb	147,000	147,000
W/S_w, lb/ft^2	86.0	86.0
Earlier criteria per CAR 4b		
K	1.24	1.24
U, fps eas	30	30
KU	36.6	36.6
Revised criteria per Amendment 4b-3		
σ	1.0	0.4481
μ_g	33.64	57.05
K_g	0.760	0.805
U_{de}, fps eas	50	46
$K_g U_{de}$	38.0	37.03

[a]The gust factor K for the earlier criteria is calculated
from Eq. (5.3), and the gust velocity is obtained from
Table 5.1; the revised criteria analysis is calculated from
Eqs. (5.5) and (5.6), and gust velocities are obtained from
Table 5.2.

database for the aircraft load factors did not change. Only the analytical approach
to the analysis was modified to derive more rational criteria using mass ratio and
a specified shape of the gust.

5.3 FAR/JAR Discrete Gust Design Criteria

The FAR/JAR harmonization working group released a copy of proposed gust
criteria on March 16, 1993. The new criteria replace the original FAR 25 discrete
gust methodology with a more rational basis that accounts for the aerodynamic
and structural dynamic characteristics of the airplane.

The discrete gust design velocities per the FAR/JAR proposal are shown in
Fig. 5.2. These gust velocities, revised from the original criteria requirements
shown in Fig. 5.1, were obtained from recorded flaps-up events in the Civil Aircraft
Airworthiness Data Recording Program (CAADRP) database from May 1980
through the end of the program in April 1990. The Boeing fleet included the
737, 757, 747-100, and 747-200 airplanes. Other aircraft were evaluated by the
manufacturers of each model used in the CAADRP database.

5.3.1 Proposed Discrete Gust Design Criteria

The discrete gust criteria as proposed for FAR/JAR 25.341(a) are as follows.

The airplane is assumed to be subjected to symmetrical vertical and lateral
gusts in level flight. Limit gust loads must be determined in accordance with the
following provisions.

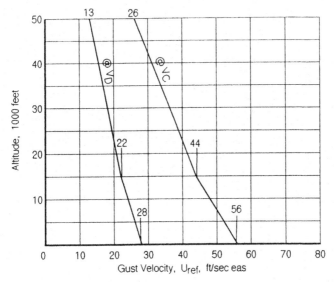

Fig. 5.2   Proposed gust velocities for discrete criteria [proposed change to FAR/JAR 25.241(a)].

1) Loads on each part of the structure must be determined by dynamic analysis. The analysis must take into account unsteady aerodynamic characteristics and all significant structural degrees of freedom including rigid-body motions.

2) The shape of the gust must be the one-minus-cosine gust as shown in Fig. 5.3 and defined by Eq. (5.8):

$$U(s) = (U_{ds}/2)\left[1 - \cos\left(\frac{\pi s}{H}\right)\right] \qquad (5.8)$$

for $(0 < s < 2H)$ and $(30 < H \leq 350)$ where

$$U_{ds} = U_{\text{ref}} F_g (H/350)^{1/6} \qquad (5.9)$$

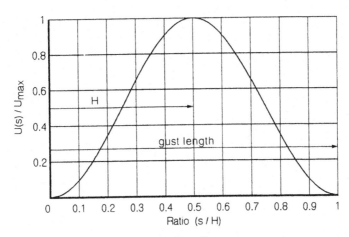

Fig. 5.3   Gust shape used for discrete gust analyses.

Table 5.4 Flight profile alleviation factors for
proposed design discrete gust criteria per
FAR/JAR 25.341 (a)[a]

$$F_g = 0.5(F_{gz} + F_{gm}) \qquad \text{at sea level} \qquad (5.9a)$$

$$F_{gz} = 1.0 - z_{mo}/250{,}000 \qquad (5.9b)$$

$$F_{gm} = [R_2 \tan(\pi R_1/4)]^{0.5} \qquad (5.9c)$$

$$R_1 = \frac{\text{maximum landing weight}}{\text{maximum takeoff weight}} \qquad (5.9d)$$

$$R_2 = \frac{\text{maximum zero fuel weight}}{\text{maximum takeoff weight}} \qquad (5.9e)$$

z_{mo} is the maximum operating altitude used for
certification, ft

[a]The flight profile alleviation factor F_g as used in Eq. (5.9) varies
linearly from the sea level value defined by Eq. (5.9a) to 1.0 at
the maximum operating altitude z_{mo}.

and where H is the gust gradient that is the distance parallel to the airplane's flight path for the gust to reach its peak velocity (ft), s is the distance penetrated into the gust (ft), U_{ref} is the reference gust velocity in equivalent airspeed defined in Fig. 5.2 for $H = 350$ ft, and F_g is the flight profile alleviation factor defined in Table 5.4.

3) A sufficient number of gust gradient distances in the range of 30–350 ft must be investigated to find the critical response for each load quantity.

When a stability augmentation system is included in the analysis, the effect of any significant system nonlinearities should be accounted for when deriving limit loads from limit gust conditions.

5.4 Continuous Turbulence Gust Loads Criteria

In 1970, Amendment 25-23 to FAR 25 incorporated FAR 25.305(d). This amendment requires the dynamic response of the airplane to vertical and lateral continuous turbulence, which must be taken into account.

Before this amendment the response to continuous turbulence was required as a special condition for several aircraft certified before 1970.

Amendment 25-54 to FAR 25 incorporated the continuous turbulence gust loads criteria, Appendix G, in September 1980 as summarized in Fig. 5.4 and Table 5.5.

Hoblit[1] has a chapter in his book on the subject of continuous turbulence criteria and the basis for determination of existing regulations.

5.4.1 Von Kármán Spectrum for PSD Analyses

The power spectral density of the atmospheric turbulence given in FAR 25, Appendix G, is the von Kármán spectrum as defined in Eq. (5.10):

$$\Phi(\Omega) = (\sigma^2 L/\pi) \frac{1 + 8/3(1.339L\Omega)^2}{[1 + (1.339L\Omega)^2]^{11/6}} \qquad (5.10)$$

where Φ is the power spectral density [(ft/s)^2/rad/ft], σ is the rms gust velocity (ft/s), Ω is the reduced frequency (rad/ft), and $L = 2500$ ft.

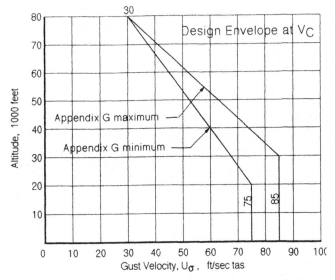

Fig. 5.4 Continuous turbulence design gust velocities (FAR 25 Appendix G).

5.4.2 Limit Design Loads

The limit design loads for continuous turbulence are determined by scaling the ratio of rms incremental load to the rms gust velocity, as shown in Eq. (5.11):

$$(\text{increment load due to gust}) = \bar{A}U_\sigma \qquad (5.11)$$

where U_σ is the gust velocity (ft/s tas), and $\bar{A}$ is the ratio of rms incremental load to rms gust velocity.

A discussion of the development of continuous turbulence gust loads criteria is found in Chapter 5 of Ref. 1; Hoblit discusses the two forms of criteria: the design envelope requirements and the mission analysis procedure and philosophy.

5.4.3 Design Gust Velocities for PSD Analyses

The design gust velocities used for continuous turbulence analyses are shown in Fig. 5.3. As stated in FAR 25, Appendix G, the minimum values may be used if the design is comparable to a similar design with extensive satisfactory service experience.

Table 5.5 Design gust velocities for continuous turbulence analyses

Design airspeeds	Gust velocity U, ft/s
V_B	$1.32U_{\sigma m}$[a]
V_C	$U_{\sigma m}$
V_D	$0.50U_{\sigma m}$

[a]Where $U_{\sigma m}$ is shown in Fig. 5.4.

The following factors must be taken into consideration in assessing comparability of a similar design. 1) The transfer functions of a new design should exhibit no unusual characteristics. 2) The typical mission of the new design is substantially equivalent to the similar design.

5.4.4 Elastic Airplane Analyses

As has been the practice over the years since the inclusion of design considerations for flight in continuous turbulence, analyses have included multiple degrees of freedom to account for the dynamic response of the aircraft structure and, where necessary, the effect of "black boxes," such as a yaw damper.

At the time of publication of this book, the FAR/JAR harmonization group had not completed the work on defining revised criteria for continuous gust design. The following is proposed for the next revision to FAR/JAR 25 whereby FAR 25.305(d) will be recoded as follows.

"FAR/JAR 25.341(b) Continuous Gust Design Criteria. The dynamic response of the airplane to vertical and lateral continuous turbulence must be taken into account. The continuous gust design criteria of FAR/JAR Appendix G must be used to establish the dynamic response unless more rational criteria are shown."

5.5   Vertical Discrete Gust Considerations

As discussed previously in this chapter, the gust design criteria have evolved in several significant ways since the early days of the forerunners of the present-day commercial jet aircraft. In this section comparison of the different criteria will be presented showing the impact these criteria changes have made on structural loads due to vertical gusts.

5.5.1   Gust Formula Approach

Since the introduction of the revised gust criteria discussed in Sec. 5.2, the traditional method in solving an encounter of an aircraft with turbulence was by using the gust formula approach for calculating vertical load factors, wing angle of attack, and horizontal tail loads.

Using the vertical load factor at the airplane center of gravity from Eq. (5.4), the variation of load factor along the airplane x axis can be obtained from Eq. (5.12):

$$n_z = n_{zcg} - \ddot{\theta} \blacktriangle x / 386.4 \qquad (5.12)$$

The pitching acceleration about the airplane center of gravity may be determined from Eq. (5.13):

$$\ddot{\theta} = (l_t / I_y)\left[\frac{(d\,BTL)}{(dn_z)}(n_{zcg} - 1) - \blacktriangle L_{tg} \right] \qquad (\text{rad/s}^2) \qquad (5.13)$$

The incremental wing angle of attack and the horizontal tail load due to a vertical gust are defined in Eqs. (5.14) and (5.15):

$$\blacktriangle \alpha_w = 57.3 K_g U_{de} / (1.69 V_e) \qquad (\text{deg}) \qquad (5.14)$$

$$\blacktriangle L_{tg} = (1 - \epsilon_{\alpha w}) \blacktriangle \alpha_w L_{t\alpha s} \qquad (\text{lb}) \qquad (5.15)$$

The gust alleviation factor K_g is obtained from Eq. (5.5). The balancing tail load parameter shown in Eq. (5.13) may be calculated from Eq. (2.22).

The method of analysis discussed in this section, identified as "gust formula method," will be used as the basis of comparison with other analyses.

5.5.2 One-DOF Discrete Vertical Gust Analysis

The current regulations do not define the transient lift functions for discrete gust analyses using the gust shapes defined by either Eq. (5.7) or Eq. (5.8). Transient lift functions, known as the Küssner and Wagner functions, are expressed differently in various reference sources. Hence the question is raised as to which set of parameters should be used in a discrete gust analysis.

The purpose of exploring the one-DOF discrete gust analysis is to show how the gust alleviation factor was obtained for the revised gust criteria. Comparison will be made using this analysis with different representations of the transient lift functions to the baseline analysis from which the revised gust criteria gust alleviation factor was obtained.

The equation of motion for an aircraft encountering a vertical gust, assuming plunge only, is developed in Ref. 5. The one-DOF equation is written as

$$\ddot{z} = (q S_w C_{L\alpha}/M)\alpha_e(s) \tag{5.16}$$

where

$$\alpha_e(s) = (U_{de}/V_e)\alpha_g(s) + \alpha_c(s) \tag{5.17}$$

and where $\alpha_g(s)$ is the effective angle of attack due to the gust defined by Eq. (5.18), and $\alpha_c(s)$ is the effective angle of attack due to damping defined by Eq. (5.21).

The equation of motion defined by Eq. (5.16) is based on the following assumptions:

1) The airplane is a rigid body.

2) The airplane forward airspeed is constant throughout the gust encounter.

3) The airplane is in steady level flight before the gust encounter.

4) The airplane is free to rise (plunge) but cannot pitch.

5) The lift increments are of the complete airplane as defined by the lift curve slope in Eq. (5.16).

6) The gust velocity is uniform across the wing span and is parallel to the vertical axis of the airplane at any instant in time.

The effective angle of attack due to a vertical gust[5] is defined as

$$\alpha_g(s) = \int \frac{1}{2\pi} C_{Lg}/(s - s_1)\frac{d[u(s_1)/U]}{ds_1}ds_1 \tag{5.18}$$

Defining the gust shape as a one-minus-cosine gust as shown in Eq. (5.8),

$$u(s)/U = [1 - \cos(2\pi s/G)]/2 \tag{5.19}$$

where s is the distance penetrated into the gust (chords), and G is the gust length (chords).

Table 5.6 Effective angle of attack due to a vertical gust

$$\alpha_g(s) = -(b_0/2)[1 - \cos(s/k)] - C_1 - C_2 - C_3 \quad (5.18a)$$

where $i = 1, 2, 3$,

$$C_i = \frac{2\pi^2 b_i[-k\beta_i \sin(s/k) + \cos(s/k) - e^{-\beta_i s}]}{(G^2\beta_i^2 + 4\pi^2)} \quad (5.18b)$$

where

$k = G/(2\pi)$ (5.18c)

$G =$ gust length, chords

$b_i =$ coefficients shown in Eq. (5.20)

$\beta_i =$ coefficients shown in Eq. (5.20)

The transient lift response due to a sharp-edge gust, the Küssner function, is defined in Eq. (5.20)[5]:

$$\frac{1}{2\pi}C_{Lg}(s) = b_0 + b_1 e^{-\beta_1 s} + b_2 e^{-\beta_2 s} + b_3 e^{-\beta_3 s} \quad (5.20)$$

where s is defined in chords (not semichords as in some references).[6]

Since the integral for the effective angle of attack due to the gust does not contain the dependent variable, it can be integrated directly. Defining the Küssner function with the coefficients as noted in Eq. (5.20), the evaluation of the integral results in the effective angle of attack due to the gust as shown in Table 5.6.

The effective angle of attack due to damping[5] is defined as

$$\alpha_c(s) = \frac{-c}{V_t} \int \frac{1}{2\pi}C_{La}(s - s_1)\frac{\ddot{z}(s)}{V_t}\,ds \quad (5.21)$$

where c is the reference mean aerodynamic chord (ft), V_t is the airplane true airspeed (ft/s), and $\ddot{z}$ is the vertical acceleration (ft/s^2).

The transient lift response due to unit change in angle of attack, the Wagner function, is defined in Eq. (5.22)[6]:

$$\frac{1}{2\pi}C_{La}(s) = b_0 + b_1 e^{-\beta_1 s} + b_2 e^{-\beta_2 s} + b_3 e^{-\beta_3 s} \quad (5.22)$$

Since $\alpha_c(s)$ includes the dependent variable z, the solution to Eq. (5.21) is not readily obtainable. Solution of the equation of motion, Eq. (5.16), may be accomplished using numerical integration by combining Eqs. (5.17) and (5.21) along with the effective angle of attack due to the gust as shown in Table 5.6.

The wing incremental angle of attack for the one-DOF analysis may be calculated from Eq. (5.17). Since this analysis assumes only the plunge degree of freedom, incremental horizontal tail loads may be calculated from Eq. (5.15):

$$\blacktriangle L_{tg} = (1 - \epsilon_{\alpha w})L_{t\alpha s}\alpha_e(s) \quad (5.23)$$

5.5.3 Two-DOF Discrete Vertical Gust Analysis

A two-DOF discrete gust analysis may be developed using the same assumptions as the one-DOF analysis, except the airplane is allowed to pitch during the gust encounter.

The equation of motion in translation may be written as

$$\ddot{z}(s) = \frac{q S_w}{M}[C_{Lato}\alpha_e(s) + c_{Lat}\alpha_t(s_t)] \tag{5.24}$$

The equation of motion in pitch may be written as

$$\ddot{\theta}(s) = \frac{q S_w c_w}{I_y}[C_{Mato}\alpha_e(s) + c_{Mat}\alpha_t(s_t)] \tag{5.25}$$

The effective angle of attack due to the gust and damping may be written as

$$\alpha_e(s) = (U_{de}/V_e)\alpha_g(s) + \alpha_c(s) \tag{5.26}$$

Equation (5.26) represents the gust encounter of the wing and body, and Eq. (5.27) represents the gust encounter with the horizontal tail:

$$\alpha_t(s_t) = (U_{de}/V_e)\alpha_g(s_t) + \alpha_c(s_t) \tag{5.27}$$

The relationship of the gust penetration at the wing and the horizontal tail is shown in Eq. (5.28):

$$s_t = s - x_t/c_w \tag{5.28}$$

The one-minus-cosine gust shape is defined by Eq. (5.19) for the gust striking the wing and by Eq. (5.29) for the gust striking the horizontal tail:

$$u(s_t)/U = [1 - \cos(2\pi s_t/G)]/2 \tag{5.29}$$

The effective angles of attack due to vertical gust are defined in a similar manner as shown in Eq. (5.18) for both the wing and horizontal tail parameters using the gust shapes defined by Eqs. (5.19) and (5.29).

In a similar manner the effective angle of attack due to damping may be obtained from Eq. (5.21). In either case the appropriate relationship as shown in Eq. (5.28) must be used in defining the Küssner and Wagner functions.

The two-DOF analysis is solved using the following equations to obtain the solution of the equations of motion using finite differences for integration:

$$\dot{\theta}_j = \dot{\theta}_{j-1} + \Delta t(\ddot{\theta}_j + \ddot{\theta}_{j-1})/2 \tag{5.30}$$

$$\dot{\alpha}_j = \dot{\theta}_j - \ddot{z}_j/V_t \tag{5.31}$$

The incremental load factor at the airplane center of gravity is obtained from Eq. (5.32):

$$\Delta n_j = \ddot{z}_j/g \tag{5.32}$$

Table 5.7 Definition of aerodynamic parameters

Tail-off pitching moment coefficient:
$$C_{M\alpha to} = C_{L\alpha to}(dC_M/dC_L + CG - 0.25) \qquad (5.25a)$$
where $C_{L\alpha to}$ is the tail-off lift curve slope, and dC_M/dC_L is the slope of the tail-off pitching moment curve.

Horizontal tail coefficients:
$$C_{L\alpha t} = L\alpha_s(1 - \epsilon_\alpha)/(qSw) \qquad (5.24a)$$
$$C_{M\alpha t} = (1 - \epsilon_\alpha)(M\alpha_s - l_t L\alpha_s)/(qSwc_w) \qquad (5.25b)$$
where $L\alpha_s$ is the lift of the horizontal tail due to stabilizer angle of attack (lb/deg), and $M\alpha_s$ is the pitching moment of the horizontal tail due to stabilizer angle of attack, about the horizontal tail pitch reference axis (in.-lb/deg).

Rate derivatives:
$$C_{L\dot\theta} = C_{L\alpha T}l_t/V_t \qquad (5.24b)$$
$$C_{L\dot\alpha} = C_{L\alpha T}x_t(1 - \epsilon_\alpha) \qquad (5.24c)$$
$$C_{M\dot\theta} = -C_{L\dot\theta}l_t/c_w \qquad (5.25c)$$
$$C_{M\dot\alpha} = -C_{L\dot\alpha}l_t/c_w \qquad (5.25d)$$

The incremental angle of attack of the wing, defined in Eq. (5.33), is obtained directly from Eq. (5.26):

$$\blacktriangle\alpha_w = \alpha_e(s) \qquad (5.33)$$

The incremental angle of attack of the horizontal tail is obtained from Eq. (5.34) where the first term is defined by Eq. (5.27):

$$\blacktriangle\alpha_t = \alpha_t(s_t) + l_t\dot\theta_j/V_t \qquad (5.34)$$

Horizontal tail incremental loads may now be calculated from Eq. (5.35) using the angle of attack of the horizontal tail as defined by Eq. (5.34):

$$\blacktriangle L_t = L\alpha_s\blacktriangle\alpha_t \qquad (5.35)$$

The definitions of the aerodynamic parameters used in these analyses are summarized in Table 5.7.

5.5.4 Example of One-DOF Discrete Vertical Gust Analyses

A discrete one-minus-cosine one-DOF (plunge) vertical gust analysis is shown in Table 5.8 assuming a gust length of 25 chords. The resulting maximum loads and angle of attack are only 1.7% lower than the gust formula results using the gust alleviation factor represented by Eq. (5.5), which was expected, since the transient lift functions were those used in the analysis from which the gust alleviation factor was derived.[5]

5.5.5 Example of Two-DOF Discrete Vertical Gust Analyses

A discrete one-minus-cosine two-DOF vertical gust analysis is shown in Table 5.9 assuming a gust length of 25 chords. Load factors computed using the gust load formula are shown for comparison purposes.

Table 5.8 Vertical gust analysis—one-DOF one-minus-cosine discrete gust[a]

	S_w, ft^2 = 1951		mac, in. = 199.7		

Alt., ft = 20,000 V_e, keas = 337.9   Mach = 0.754 V_t, fps = 782.3

Gross weight, lb = 252,000 CG, % mac/100 = 0.204

U_{de}, fps eas = 50   Grad., ft = 208.02   Gust length = 25 chords

Gust formula data:

K_g = 0.811 μ_g = 62.59

$\Delta\alpha_w$, deg = 4.070 $n_z - 1$ = 1.295 $\blacktriangle L_{tg}$, lb = 26,462

Integration increments = 160

I	s, chords	Time, s	$\Delta\alpha_w$, deg	$n_z - 1$	$\blacktriangle L_t$, lb
70	10.938	0.233	3.626	1.154	23,573
71	11.094	0.236	3.670	1.168	23,861
72	11.250	0.239	3.712	1.181	24,132
73	11.406	0.243	3.751	1.193	24,387
74	11.563	0.246	3.787	1.205	24,625
75	11.719	0.249	3.821	1.216	24,845
76	11.875	0.253	3.852	1.226	25,046
77	12.031	0.256	3.881	1.235	25,230
78	12.188	0.259	3.906	1.243	25,395
79	12.344	0.263	3.928	1.250	25,541
80	12.500	0.266	3.948	1.256	25,667
81	12.656	0.269	3.964	1.261	25,775
82	12.813	0.273	3.978	1.266	25,863
83	12.969	0.276	3.988	1.269	25,931
84	13.125	0.279	3.996	1.271	25,979
85	13.281	0.283	4.000	1.273	26,007
86	13.438	0.286	4.001	1.273	26,015
87	13.594	0.289	4.000	1.273	26,003
88	13.750	0.292	3.995	1.271	25,971
89	13.906	0.296	3.987	1.268	25,919
90	14.063	0.299	3.976	1.265	25,847
91	14.219	0.302	3.961	1.260	25,755
92	14.375	0.306	3.944	1.255	25,643
93	14.531	0.309	3.924	1.248	25,512
94	14.688	0.312	3.901	1.241	25,361
95	14.844	0.316	3.875	1.233	25,191

[a]The gust shape is defined by Eq. (5.19) assuming a gust length of 25 chords. The Küssner and Wagner functions are as defined in Ref. 5. Wing angle of attack and horizontal tail loads are calculated using Eqs. (5.17) and (5.23).

Table 5.9a **Vertical gust analysis—two-DOF one-minus-cosine discrete gust[a]**
(summary of maximum accelerations and tail loads)

Alt., ft = 20,000		V_e, keas = 337.9	Mach = 0.754		V_t, fps = 782.3
Gross weight, lb = 252,000			CG, % mac/100 = 0.204		
U_{de}, fps eas = 50		Grad., ft = 208.02	Gust length = 25 chords		

I	s, chords	Time, s	$\Delta\alpha_w$, deg	$n_z - 1$	$\ddot\theta$, rad/s^2	$\blacktriangle L_t$, lb
		Time of maximum load factor at airplane CG				
87	13.594	0.289	4.027	1.218	−0.084	19,125
		At horizontal tail		1.1396		
		Time of maximum pitching acceleration				
114	17.813	0.379	2.881	0.890	−0.122	22,185
		At horizontal tail		1.152		
		Time of horizontal tail maximum load				
107	16.719	0.356	3.355	1.030	−0.119	22,719
		At horizontal tail		1.284		
		Time of maximum load factor at stabilizer pivot axis				
92	14.375	0.306	3.969	1.204	−0.095	20,656
		At horizontal tail		1.408		

Gust formula data
$K_g = 0.811$ $\mu = 62.59$ $\Delta\alpha_w$, deg = 4.070
$n_z - 1 = 1.295$ Pitch acc., rad/s^2 = −0.286 $\blacktriangle L_{tg}$, lb = 26,462
$n_z - 1$ at horizontal tail = 1.907
Gust frequency vs short period pitch frequency:
    Gust freq., rad/s = 0.940
    Pitch freq., rad/s = 1.958
    Chords for pitch freq. = 12.0

[a]The gust shape is defined by Eq. (5.19) assuming a gust length of 25 chords. The Küssner and Wagner functions are as defined in Ref. 5. Wing and horizontal tail angle of attack are calculated using Eqs. (5.26) and (5.27). Horizontal tail loads are calculated using Eq. (5.35). The aerodynamic parameters used for this analysis are summarized in Table 5.10.

The time history analysis shows about a 6% reduction in incremental load factor at the airplane center of gravity, whereas the reduction in load factor at the horizontal tail is considerable, like 26% from the gust formula data. The effect of the gust encounter at the wing vs the time of the gust encounter with the horizontal tail results in a significant reduction in airplane pitching acceleration, hence the reduction in load factor at the horizontal tail over the gust formula.

5.5.6 Effect of Gust Gradient

The effect of gust gradient on load factors obtained from the two-DOF analysis is shown in Table 5.10 for a series of conditions from 12 to 42 chords (350 ft for the aircraft used). The gust velocity is assumed constant.

Table 5.9b Vertical gust analysis—two-DOF one-minus-cosine discrete gust (time history)

Alt., ft = 20,000		V_e, keas = 337.9		Mach = 0.754	V_t, fps = 782.3
Gross weight, lb = 252,000				CG, % mac/100 = 0.204	
U_{de}, fps eas = 50		Grad., ft = 208.02		Gust length = 25 chords	

I	s, chords	Time, s	$\Delta\alpha_w$, deg	$n_z - 1$	$\ddot{\theta}$, rad/s^2	$\blacktriangle L_t$, lb
60	9.375	0.199	3.080	0.916	−0.014	7,531
61	9.531	0.203	3.145	0.936	−0.116	7,964
62	9.688	0.206	3.209	0.956	−0.119	8,402
63	9.844	0.209	3.272	0.975	−0.021	8,845
64	10.000	0.213	3.332	0.993	−0.023	9,293
65	10.156	0.216	3.391	1.012	−0.026	9,745
66	10.313	0.219	3.447	1.029	−0.028	10,199
67	10.469	0.223	3.501	1.046	−0.031	20,656
68	10.625	0.226	3.553	1.062	−0.033	11,115
69	10.781	0.229	3.603	1.077	−0.036	11,574
70	10.938	0.233	3.650	1.092	−0.038	12,034
71	11.094	0.236	3.695	1.106	−0.041	12,493
72	11.250	0.239	3.737	1.120	−0.044	12,950
73	11.406	0.243	3.777	0.132	−0.046	13,405
74	11.563	0.246	3.814	0.144	−0.049	13,857
75	11.719	0.249	3.848	0.155	−0.052	14,305
76	11.875	0.253	3.879	0.165	−0.055	14,188
77	12.031	0.256	3.908	1.174	−0.057	15,188
78	12.188	0.259	3.933	1.183	−0.060	15,620
79	12.344	0.263	3.956	1.190	−0.063	16,046
80	12.500	0.266	3.976	1.197	−0.065	16,464
81	12.656	0.269	3.992	1.203	−0.071	17,875
82	12.813	0.273	4.006	1.207	−0.073	17,276
83	12.969	0.276	4.016	1.211	−0.076	18,668
84	13.125	0.279	4.024	1.214	−0.076	18,049
85	13.281	0.283	4.028	1.216	−0.079	18,420
86	13.438	0.286	4.029	1.217	−0.081	18,779
87	13.594	0.289	4.027	1.218	−0.084	19,125
88	13.750	0.292	4.022	1.217	−0.086	19,459
89	13.906	0.296	4.013	1.215	−0.088	19,780
90	14.063	0.299	4.002	1.212	−0.091	20,087
91	14.219	0.302	4.987	1.209	−0.093	20,379
92	14.375	0.306	3.969	1.204	−0.095	20,656
93	14.531	0.309	3.949	1.199	−0.097	20,918
94	14.688	0.312	3.925	1.192	−0.100	21,163
95	14.844	0.316	3.898	1.185	−0.102	21,392

Table 5.9b   Vertical gust analysis—two-DOF one-minus-cosine discrete gust
(time history) (continued)

I	s, chords	Time, s	$\Delta\alpha_w$, deg	$n_z - 1$	$\ddot{\theta}$, rad/s^2	$\blacktriangle L_t$, lb
96	15.000	0.319	3.868	1.177	−0.104	21,604
97	15.156	0.322	3.835	1.167	−0.105	21,799
98	15.313	0.326	3.799	1.157	−0.107	21,976
99	15.469	0.329	3.760	1.146	−0.109	22,135
100	15.625	0.332	3.719	1.135	−0.110	22,275
101	15.781	0.336	3.675	1.122	−0.112	22,397
102	15.938	0.339	3.628	1.109	−0.113	22,499
103	16.094	0.342	3.578	1.094	−0.115	22,583
104	16.250	0.346	3.526	1.079	−0.116	22,647
105	16.406	0.349	3.471	1.063	−0.117	22,691
106	16.563	0.352	3.414	1.047	−0.118	22,715
107	16.719	0.356	3.355	1.030	−0.119	22,719
108	16.875	0.359	3.293	1.012	−0.120	22,704
109	17.031	0.362	3.230	0.993	−0.120	22,668
110	17.188	0.366	3.164	0.974	−0.121	22,612
111	17.344	0.369	3.096	0.954	−0.121	22,535
112	17.500	0.372	3.026	0.933	−0.122	22,439
113	17.656	0.376	3.955	0.912	−0.122	22,322
114	17.813	0.379	3.881	0.890	−0.122	22,185
115	17.969	0.382	3.807	1.868	−0.122	22,029
116	18.125	0.386	2.731	0.846	−0.122	21,852
117	18.281	0.389	2.653	0.823	−0.122	21,656
118	18.438	0.392	2.574	0.799	−0.121	21,440
119	18.594	0.396	2.494	0.775	−0.121	21,205
120	18.750	0.399	2.413	0.751	−0.120	20,952
121	18.906	0.402	2.331	0.727	−0.120	20,679
122	19.063	0.405	2.249	0.702	−0.119	20,388
123	19.219	0.409	2.165	0.677	−0.118	20,079
124	19.375	0.412	2.081	0.652	−0.117	19,752
125	19.531	0.415	1.997	0.627	−0.116	19,408

The most significant effect of gust gradient in this analysis is the load factor at the horizontal tail. Depending on the airplane pitch characteristics, other gradients may be critical with respect to a 25-chord gust.

5.6 Transient Lift Functions

Consideration must be given to the transient lift functions used in the discrete gust analyses, whether single DOF or multiple DOF (including structural response).

The transient lift functions, commonly called the Küssner and Wagner functions, are defined in Table 5.11 for some of the more commonly used representations of these functions. The coefficients of Eqs. (5.20) and (5.22) are normalized to a value of unity for b_0 in Table 5.11.

Table 5.10   Response parameters obtained from two-DOF one-minus-cosine discrete vertical gust analysis[a]

Cond. code[b]	U_{de}, ft/s eas	Gust length, chords	Gust grad., ft	$\blacktriangle n_z$ at CG	$\blacktriangle n_z$ at horizontal tail
1	50	12	99.95	1.125	1.153
				1.088	1.187
1	50	15	124.81	1.167	1.257
				1.135	1.291
1	50	20	166.42	1.203	1.357
				1.186	1.377
1	50	25	208.02	1.218	1.396
				1.204	1.408
1	50	30	249.63	1.219	1.405
				1.211	1.413
1	50	35	291.23	1.213	1.397
				1.210	1.404
1	50	42.06	350.0	1.197	1.375
				1.196	1.377
3	50	25	208.02	1.253	1.435
				1.240	1.448
Gust formula	50	—	—	1.295	1.907

[a]The gust shape is as defined by Eqs. (5.19) and (5.29) with a variable gust length as noted in this table. Load factors are shown at the time of maximum load factor at the airplane center of gravity and the time of maximum load factor at the horizontal tail.
[b]See Table 5.11.

The representation of the Küssner and Wagner functions shown in Ref. 5 (condition code 1) is normally used for vertical discrete gust analyses since the design gust velocities in FAR 25.341(a) were derived from analyses based on these coefficients. The variation of these functions with gust penetration is shown in Table 5.12.

5.6.1   Transient Lift Function Effect on Response Parameters of One-DOF Analysis

The effects of various representations of the Küssner and Wagner functions on the response to a discrete gust using the one-DOF analysis are summarized in Table 5.13. There is good agreement between the baseline analysis and the gust formula results, which was the basis for the empirical derivation of the gust alleviation factor K_g.

Table 5.11   Transient lift Küssner and Wagner functions

Cond. code[a]	AR	Mach	Φ, s	b_0	b_1	b_2	b_3	β_1	β_2	β_3
1[5]	∞	0	KF[b]	1.0	−0.236	−0.513	−0.171	0.116	0.728	4.84
			WF[b]	1.0	−0.165	−0.335	0	0.090	0.600	0
2*[6]	6	0	KF	1.0	−0.448	−0.272	−0.193	0.580	1.45	6.0
			WF	1.0	−0.361	0	0	0.762	0	0
3*[7]	∞	0	KF	1.0	−0.500	−0.500	0	0.260	2.00	0
			WF	1.0	−0.165	−0.335	0	0.0910	0.60	0
4*[6]	∞	0.70	KF	1.0	−0.402	−0.461	−0.137	0.1084	0.625	2.948
			WF	1.0	−0.364	−0.405	0.419	0.1072	0.714	1.804

[a]The functions obtained from the references noted by * are changed to s of chords rather than the semichords shown in the reference. The data shown have been normalized to an asymptotic value of unity.
[b]KF = Küssner functions, Eq. (5.20), and WF = Wagner function, Eq. (5.22).

Table 5.12   Küssner and Wagner functions
used for the baseline analyses, variation
with gust penetration,[a] with aspect ratio =
infinite, Mach = 0, and gust length =
25 chords

s, chords	Time,[b] s	Küssner function	Wagner function
0	0	0.080	0.500
2.5	0.053	0.740	0.793
5.0	0.106	0.854	0.878
7.5	0.160	0.899	0.912
10.0	0.213	0.926	0.932
12.5	0.266	0.945	0.946
15.0	0.319	0.959	0.957
17.5	0.372	0.969	0.966
20.0	0.425	0.977	0.973
22.5	0.479	0.983	0.978
25.0	0.532	0.987	0.983

[a]The Küssner and Wagner functions as compiled in this table were calculated using the definitions for condition code 1 as shown in Table 5.11.
[b]Time, although not required for these calculations, is shown for reference purposes for the analysis given in Table 5.8.

Table 5.13 Effect of various definitions of the Küssner and Wagner functions on airplane gust response for a one-DOF discrete gust analysis[a]

Cond. code[b]	s at max $\blacktriangle n_z$, chords	Time, s	$\blacktriangle \alpha_w$, deg	$\blacktriangle n_z$, airplane CG
1[c]	13.438	0.286	4.001	1.273
2	12.969	0.276	4.420	1.406
3	13.594	0.289	4.117	1.310
4	14.063	0.299	3.670	1.168
Gust formula			4.070	1.295

[a]The gust shape is defined by Eq. (5.19) assuming the gust length as 25 chords. The analysis for condition 1 is shown in Table 5.8. All other condition parameters are the same except for the Küssner and Wagner functions.
[b]See Table 5.12 for identification of the Küssner and Wagner functions used for these analyses.
[c]Baseline analysis is shown in Table 5.8.

Good agreement is also obtained for the analysis using the transient functions obtained from Fung.[7] Analyses using transient functions for infinite aspect ratio and zero Mach number are more in agreement than analyses using either aspect ratio 6 or Mach 0.70 data. Since the lift curve slopes already include aspect ratio and compressibility effects, the general practice is to use the transient lift functions for infinite aspect ratio and Mach zero.

5.7 Vertical Gust Continuous Turbulence Considerations

Much discussion and work has been accomplished in developing the analysis techniques and methodology for aircraft encountering continuous turbulence. The regulations as noted in Sec. 5.4 require an analysis for continuous turbulence. The resulting gust loads include multiple DOF, thus incorporating both rigid-body motion and the flexible structural modes representing the elastic deformation of the structure.

The basic concepts for the development of continuous turbulence analyses are discussed in Refs. 1 and 6–8. A system for solution of the resulting equations of motion is given in Ref. 9.

5.7.1 PSD Vertical Gust Formula

Consideration will be given to the PSD vertical gust formula developed by Houbolt,[8] whereby he derives an equation for continuous turbulence analogous to the discrete gust formula, Eq. (5.4). The gust alleviation factor K_g is replaced by the spectral gust alleviation factor K_Φ, and the derived discrete gust velocity U_{de} in feet per second equivalent airspeed is replaced by the gust velocity U_σ in feet per second true airspeed.

Inserting Eqs. (5.36) and (5.37) into Eq. (5.4) and exchanging K_Φ for K_g, one may derive the vertical gust formula PSD equation:

$$V_e = V_t \sigma^{\frac{1}{2}} / (1.688) \qquad \text{(keas)} \qquad (5.36)$$

$$U_{de} = U_t \sigma^{\frac{1}{2}} \qquad \text{(ft/s equivalent airspeed)} \qquad (5.37)$$

where V_t is the true airspeed (ft/s), and U_t is the true gust velocity (ft/s).

The PSD vertical gust formula as derived by Houbolt[8] becomes

$$\blacktriangle n_z = K_\Phi V_t \rho S_w C_{L\alpha} U_\sigma / (2W) \qquad (5.38)$$

The spectral gust alleviation factor K_Φ may be calculated as shown in Ref. 8 using Eq. (5.39):

$$K_\Phi = \left[\int_0^{k_c} \phi_1(k) \, dk \right]^{\frac{1}{2}} \qquad (5.39)$$

where

$$\phi_1(k) = f_1(k) f_2(k) (\phi_w / \sigma_w^2) \qquad (5.40)$$

$$f_1(k) = 16\mu^2 / [(2F/k)^2 + (4\mu + 2G/k)^2] \qquad (5.41)$$

$$f_2(k) = 1/(1 + 4.92k + 2.06k^2) \qquad (5.42)$$

where k is the reduced frequency, μ is the mass ratio [defined by Eq. (5.6)], k_c is the cutoff frequency, and F and G are coefficients of Theodorsen's function (defined on page 214 of Ref. 7).

The gust velocity power spectral density used in Eq. (5.43) is modified from Eq. (5.10) to agree with the derivation in Ref. 8:

$$\frac{\phi_w}{\sigma_w^2} = \left(\frac{2L}{\pi c} \right) \frac{1 + 8/3(1.339L\Omega)^2}{[1 + (1.339L\Omega)^2]^{11/6}} \qquad (5.43)$$

5.7.2 Use of PSD Vertical Gust Formula

The question arises as to what use is the PSD vertical gust formula as defined by Eq. (5.38). The following possibilities should be considered.

1) The simplified procedure could be used to evaluate the criticality of a condition at one altitude vs another. This would be used as an aid in selecting the conditions for which a full dynamic analysis will be accomplished.

2) The simplified equation could be used to assess design U_σ values similar to the selection of the original discrete gust velocities that were based on the revised gust formula of Eq. (5.4).

Some interesting results may be obtained by calculating the U_σ required to produce the same incremental load factors as the discrete gust load formula using the FAR reference gust velocities shown in Fig. 5.2. Gust load factors are calculated in Table 5.14 using Eq. (5.38). These results are computed using K_Φ based on the von Kármán description of the gust spectra with a scale of turbulence, $L = 2500$ ft.

Table 5.14 Continuous turbulence vertical gust analysis using
the gust formula approach for one of the aircraft summarized
in Fig. 5.5[a]

Alt., ft	0	15,000	20,000	23,230	27,100
V_e, keas	350.0	341.8	337.9	335.0	330.9
Mach	0.529	0.688	0.754	0.800	0.860
Gross weight, lb	206,400	206,400	206,400	206,400	206,400
CG, % mac/100	0.097	0.097	0.097	0.097	0.097
$C_{L\alpha to}$, per deg	0.08071	0.08743	0.09642	0.10284	0.10195
$(1 - \epsilon_{\alpha w})$, per deg	0.550	0.525	0.512	0.502	0.457
$C_{L\alpha w}$, per deg	0.08856	0.09548	0.10482	0.11150	0.11019
μ_g	32.72	48.23	51.88	54.50	63.24
Discrete gust analysis:					
K_g	0.757	0.793	0.798	0.802	0.812
F_g	0.810	0.878	0.900	0.915	0.933
U_{de}, fps eas	45.35	38.62	37.30	36.39	35.23
Δn_z	1.158	1.087	1.147	1.186	1.134
PSD gust analysis[b] ($L = 2500$ ft):					
Integration cutoff frequency, $k = 1$					
K_ϕ	0.494	0.553	0.564	0.570	0.589
$\bar{A}$	0.01665	0.01558	0.01585	0.01600	0.01507
μ_σ, fps tas	69.54	69.77	72.39	74.10	75.27
Δn_z	1.158	1.087	1.147	1.186	1.134

[a]The discrete gust analysis load factors are calculated using Eq. (5.4). The continuous turbulence U_σ is computed for the same load factors as obtained from the discrete gust analysis. Continuous turbulence analysis load factors are calculated using Eq. (5.38).
[b]The gust velocity power spectral density is the von Kármán spectrum defined by Eq. (5.43).

An analysis was run for five different jet transport aircraft at their respective V_{mo} speeds, and the U_σ was computed to produce the same incremental load factor as the discrete gust load factor formula at five altitudes. These data are summarized in Fig. 5.5 as a function of wing area.

The results of this study show a variation of U_σ with airplane size that is similar to the results that are concluded by Houlbolt. He states in Ref. 8, "There is still something mysterious about the scale of turbulence L, and the response parameters ... for the various aircraft."

The program used in this analysis calculates K_Φ using the Theodorsen functions as published on page 214 of Fung.[7] The integration cutoff frequency k_c, in Eq. (5.39) was selected as $k_c = 1$, and the scale of turbulence was 2500 ft.

5.8 Multiple DOF Analyses

The proposed regulations as discussed in Sec. 5.3.1 state that the analysis not only must take into account unsteady aerodynamic characteristics but also must

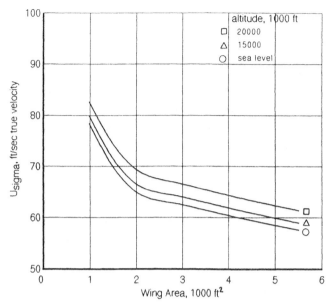

Fig. 5.5 **Usigma vs airplane wing area for PSD analysis at load factors for discrete gust analysis. The Usigma values are calculated using the PSD gust formula method defined in Eq. (5.38); the load factors are assumed the same calculated from the gust formula of Eq. (5.4) with a U_{de} of the same magnitude of U_{ds}. (See Table 5.14).**

include all significant structural degrees of freedom along with the rigid-body motions.

For aircraft having an autopilot or active controls that may enhance aerodynamic stability or other flight characteristics, the control laws must be included in the analysis to ascertain the effect on gust loads.

For the early jets the problem of the adequacy of the number of significant modes used in a dynamic analysis was of a concern due to the lack of computing capability. Now that analysis techniques have grown in sophistication and modern digital computers with very large capacities have become available, a large number of significant structural degrees of freedom can be included in the analysis.

Among the many references that already cover the subject of multiple DOF continuous turbulence and discrete gust analyses,[1,6,7] Ref. 9 presents a system of equations and solutions of the resulting aircraft response using a program called Dyloflex.

The difficulty of providing loads that are adequate for stress analysis is discussed in Ref. 1. The problem for the discrete gust analysis loads is the number of conditions available for analysis. The difficulty is how the continuous turbulence loads are to be used by the stress analyst considering that finite element analyses require a balanced load solution throughout the structure that is being modeled.

5.9 Lateral Gust Considerations

The lateral gust considerations have historically involved the lateral gust formula and a full dynamic analysis, including structural degrees of freedom and automatic flight controls such as a yaw damper.

5.9.1 Lateral Gust Load Formula (Historical Perspective)

Effective March 13, 1956, the lateral gust requirements of CAR $4b^2$ were changed by Amendment 4b-3 to incorporate the lateral gust load formula, with the gust alleviation factor, defined by Eq. (5.5), based on the lateral mass ratio.

Since the lateral gust formula is primarily related to the vertical tail, reference should be made to Sec. 9.4 in Chapter 9.

Amendment 4b-3 also incorporated the discrete gust shape that was used for lateral gust dynamic analyses of the early jet transports until the inclusion of continuous turbulence requirements in FAR 25.305(d)[10] by Amendment 25-23 in 1970. Amendment 25-54 incorporated Appendix G requirements in 1980.

5.9.2 Multiple DOF Analyses

The regulations, as proposed by the FAR/JAR harmonization process, will require that the analyses not only must take into account unsteady aerodynamic characteristics but also must include significant structural degrees of freedom along with the rigid-body motions.

For models having active controls, such as a yaw damper, the control laws must be included in the analysis to ascertain the effect on gust loads.

The equations of motion and the related load equations for solving the lateral gust analysis for either the continuous turbulence (PSD) or the discrete gust requirements are discussed in Ref. 9. Active controls, such as a yaw damper, are included in the equations of motion.

5.9.3 Transient Lift Functions for Lateral Gust Analyses

The Küssner and Wagner functions for lateral gust analyses are usually based on an aspect ratio of 3.0. These values, traditionally used for dynamic analyses, are assumed to be more representative of the vertical tail aspect ratio than the infinite aspect ratio values from which the revised discrete gust velocities were derived.

5.10 Oblique Gusts

The commercial regulations in the United States did not recognize specific combined gust criteria other than as stated in FAR 25.427.

FAR 25.427 Unsymmetrical Loads:

(b)(2) For empennage arrangements where the horizontal tail surfaces have appreciable dihedral or are supported by the vertical tail surfaces, the surfaces and supporting structure must be designed for the combined vertical and horizontal surface loads resulting from each prescribed flight load condition separately.

During the FAR/JAR harmonization process the inclusion of the so-called "around-the-clock" gust criteria was proposed as follows.

FAR/JAR 25.427(c) For empennage arrangements where the horizontal tail surfaces have dihedral angles greater than 10 degrees, or are supported by the vertical tail surfaces, the surfaces and supporting structure must be designed for gust velocities specified in 25.341(a) acting in any orientation at right angles to the flight path.

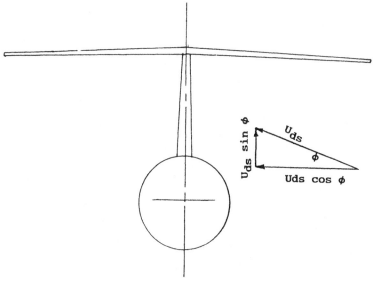

Fig. 5.6   Oblique gust.

The around-the-clock gust requirements shown in Fig. 5.6 use the discrete gust U_{ref} values as shown in Fig. 5.2.

5.10.1   Historical Perspective

The 727 series airplanes due to the "T" tail empennage arrangement did include combined load effects in the design of the horizontal tail and supporting structure. The tail support and vertical fin were critical due to the combined load effect of lateral gusts acting on the vertical tail and the unsymmetrical loads acting on the horizontal tail due to lateral gusts. Oblique around-the-clock gusts were not considered for design.

Some of the later aircraft designed after 1980 were required by the European certifying agencies to design for the around-the-clock gusts, even though these aircraft had horizontal tail dihedrals less than 10 deg.

In general the aft fuselage monocoque structure and aft body empennage support structure were critical for these load requirements.

5.10.2   Component Loads Due to Oblique Gusts

The lateral and vertical component of the oblique gust may be calculated from lateral and vertical gust design conditions by the following procedure:

$$L_{\text{lat}} = (U_{\text{lat}}/U_{ds})(L_{\text{lat des}}) \qquad (5.44)$$

where L_{lat} represents the loads on the horizontal tail and vertical tail due to lateral gust of a magnitude U_{lat},

$$L_{\text{ver}} = (U_{\text{ver}}/U_{ds})(L_{\text{ver des}} - L_{1g}) + L_{1g} \qquad (5.45)$$

where L_{ver} represents the loads on the horizontal tail and vertical tail due to vertical gust of a magnitude U_{ver}.

The vertical and lateral components of the oblique gust are determined using Eqs. (5.46) and (5.47):

$$U_{lat} = U_{ds} \cos \phi \qquad (5.46)$$

$$U_{ver} = U_{ds} \sin \phi \qquad (5.47)$$

where ϕ is the angle of oblique gust as shown in Fig. 5.6, L_{1g} represents the loads on the horizontal and vertical tail due to 1-g flight, $L_{lat\ des}$ represents the design loads due to lateral gust at U_{ds}, and $L_{ver\ des}$ represents the design loads due to vertical gust at U_{ds}.

The critical structure may now be analyzed by the stress engineers in the manner in which they are accustomed; i.e., applying lateral and vertical loads to the structure as is done in the lateral gust conditions in which the gust is horizontal and perpendicular to the airplane flight path.

Calculations are shown in Table 5.15, based on loads using the gust formula to show the method of analysis represented by Eqs. (5.44) and (5.45).

The combination of vertical and lateral load conditions based on dynamic discrete analyses may reflect time correlation of the gust loads to reduce undue conservatisms.

5.10.3   Word of Caution Concerning Oblique Gust Analyses

Some engineers have determined the combined loads by taking the square root of the sum of the squares of the vertical and lateral design conditions; i.e.,

$$\text{combined load} = \left[(L_{lat\ des})^2 + (L_{ver\ des})^2 \right]^{\frac{1}{2}} \qquad (5.48)$$

The problem with this approach is that the engineer is effectively applying an oblique gust greater than is required by the regulations for the around-the-clock

Table 5.15   Example of oblique gust analysis based on loads using the gust formula results to show the method of analysis that may be used in determining oblique gust loads[a]

ϕ, deg	U_{lat}, ft/s eas[b]	U_{ver}, ft/s eas[c]	L_{lat}, lb limit[d]	L_{ver}, lb limit[e]
0	50	0	32,833	−19,815
30	43.3	−25.0	28,434	−32,671
45	35.4	−35.4	23,246	−38,018
60	25.0	−43.3	16,417	−42,081
90	0	−50	0	−45,526

[a]The loads shown are based on U_{de} gust values for a constant speed–altitude condition where altitude = 20,000 ft, V_e = 337.9 keas, Mach = 0.754 and an assumption that $U_{ds} = U_{de} = 50$ ft/s eas.
[b]See Eq. (5.46).
[c]See Eq. (5.47).
[d]See Eq. (5.44).
[e]See Eq. (5.45).

Table 5.16   Example of head-on gust analysis[a]

Flap position	V_f, keas	$V_f + 15$, keas	n_z[b]
30	175	190	1.18
20	202	217	1.15
5	223	238	1.14

[a]The vertical load factor can be calculated from Eq. (5.52), assuming the initial load factor is 1.0 at placard speed V_f; the variation of incremental load factor is shown for a head-on gust of 25 fps (15 keas).
[b]See Eq. (5.52).

gust condition. The approach is conservative but may cause undue weight or unnecessary structural changes. For the oblique gust example shown in Table 5.16 the conservatism, if the approach of Eq. (5.48) were used, would be a factor of 1.26.

5.11   Head-On Gusts

The head-on gust requirements of FAR/JAR 25 are primarily of concern in establishing the structural design airspeeds and the structural design loads on the wing high-lift devices.

5.11.1   High-Lift Devices

In general, wing high-lift devices are critical for the trailing-edge flap loads at placard plus head-on gust, whereas the leading-edge devices are maneuver critical.

Per the requirements of FAR/JAR 25.345(b)(2), the structure must be designed for a head-on gust of 25 fps (eas). The en route requirements stated in 25.345(c) do not specifically require a gust velocity higher than that required by 25.345(b)(2).

The general analysis method is to assume that the head-on gust causes a change in dynamic pressure q with no alleviation in the resulting loads due to the time of application of the gust (as is done in the lateral or vertical gust analyses).

5.11.2   Airplane Load Factors Due to Head-On Gust at Flap Placard Speeds

The initial level flight (1-g) condition may be defined as shown in Eq. (5.49):

$$W = C_{L1} q_1 S_w \qquad (5.49)$$

where $q_1 = V_F^2/295$ (lb/ft^2), V_F is the flap placard speed (keas), C_{L1} is the lift coefficient at placard speed, and S_w is the wing reference area (ft^2).

Assuming a sudden head-on gust with no change in airplane pitch attitude during the gust occurrence (no change in wing angle of attack), the vertical load factor may be related to lift using Eq. (5.50):

$$n_z W = C_{L1} q_2 S_w \qquad (5.50)$$

where n_z is the vertical load factor due to head-on gust, and q_2 is the dynamic pressure due to the head-on gust plus initial velocity (lb/ft^2).

Assuming a head-on gust of the magnitude U_g, then one can calculate the dynamic pressure due to the gust increase from Eq. (5.51):

$$q_2 = (V_F + U_g)^2/295 \qquad \text{(lb/ft}^2) \qquad (5.51)$$

Combining Eqs. (5.49–5.51), one can calculate the vertical load factor from Eq. (5.52):

$$n_z = q_2/q_1 = (V_F + U_g)^2/V_F^2 \qquad (5.52)$$

Load factors are calculated for three flap positions in Table 5.16 using Eq. (5.52).

5.11.3 Head-On Gust and Structural Design Airspeeds

The effect of head-on gusts on the determination of the structural design airspeeds, VC and VD, are discussed in Chapter 14.

References

[1]Hoblit, F. M., *Gust Loads on Aircraft: Concepts and Applications,* AIAA Education Series, AIAA, Washington, DC, 1988.

[2]Anon., "Part 4b—Airplane Airworthiness Transport Categories," Civil Air Regulations, Civil Aeronautics Board, Dec. 1953.

[3]Donely, P., "Summary of Information Relating to Gust Loads on Aircraft," NACA Rept. 997, 1950.

[4]Pratt, K. G., "A Revised Formula for Calculation of Gust Loads," NACA TN 2964, June 1953.

[5]Pratt, K. G., and Walker, W. G., "Revised Gust Load Formula and a Re-Evaluation of V-G Data Taken on Civil Transport Airplanes from 1933 to 1950," NACA Rept. 1206, Sept. 1953.

[6]Bisplinghoff, R. L., Ashley, H., and Halfman, R. L., *Aeroelasticity,* Addison–Wesley, Reading, MA, 1955.

[7]Fung, Y. C., *An Introduction to the Theory of Aeroelasticity,* Wiley, New York, 1955.

[8]Houbolt, J. C., "Design Manual for Vertical Gusts Based on Power Spectral Techniques," Air Force Flight Dynamics Lab, AFFDL-TR-70-106, Wright-Patterson AFB, OH, Dec. 1970.

[9]Miller, R. D., Kroll, R. I., and Clemmons, R. E., "Dynamic Loads Analysis System (Dyloflex) Summary," NASA CR-2846, May 1978.

[10]Anon., "Part 25—Airworthiness Standards: Transport Category Airplanes," Federal Aviation Regulations, U.S. Dept. of Transportation, Jan. 1994.

6
Landing Loads

The landing load requirements for commercial aircraft as set forth in FAR/JAR 25.473 have not changed significantly since the design of the early jet aircraft certified in 1958/59, except for the inclusion of specific wording related to structural dynamic response in calculating airplane loads.

In general the landing loads in this chapter will cover the following subjects: 1) criteria, 2) landing speeds, 3) two-point landing conditions, 4) three-point landing conditions, 5) one-gear landing conditions, 6) side load conditions, 7) rebound landing conditions, 8) nose gear pitchover considerations, 9) landing gear shock absorption and drop tests, and 10) automatic ground spoiler considerations.

6.1 Criteria per FAR/JAR 25.473

The following criteria incorporate the proposed changes made in February 1993 as part of the FAR/JAR harmonization process.

1) For the landing conditions specified in paragraphs 25.479–25.485 the airplane is assumed to contact the ground a) in the attitudes defined in 25.479 for level landing conditions and 25.481 for taildown conditions, b) with a limit descent velocity of 10 fps at the design landing weight, and c) with a limit descent velocity of 6 fps at the design takeoff weight; d) the prescribed descent velocities may be modified if it is shown that the airplane has design features that make it impossible to develop these velocities.

2) Airplane lift, not exceeding airplane weight, may be assumed unless the presence of systems or procedures significantly affects the lift.

3) The method of analysis of airplane and landing gear loads must take into account at least the following elements: a) landing gear dynamic characteristics, b) spin up and spring back, c) the rigid-body response, and d) the structural dynamic response of the airframe if significant.

4) The limit inertia factors corresponding to the required limit descent velocities must be validated by tests as defined in 25.723(a).

5) The coefficient of friction between the tires and the ground may be established by considering the effects of skidding velocity and tire pressure. However, this coefficient of friction need not be more than 0.8.

6.2 Landing Speed Calculations

Landing speeds per FAR 25.479(a) are specified to bound the possible operation of the aircraft as follows:

$$V_{\text{landing sd}} = V L_1 = V_{so} \qquad \text{(tas)} \qquad (6.1)$$

where V_{so} is the airplane stall speed for a standard day operation at sea level with the flaps in the appropriate landing configuration,

$$V_{\text{landing hd}} = 1.25VL_2 + V_{\text{tw}} \qquad \text{(tas)} \qquad (6.2)$$

where VL_2 is the airplane stall speed for a hot day operation at 41°F above standard at the maximum airport altitude for which the airplane will be certified, and where $V_{\text{tw}} = 0$ for aircraft certified for tail winds of 10 kn or less, and V_{tw} is the tail wind when the airplane is to be certified for tail winds greater than 10 kn.

6.2.1 Effect of Hot Day on Landing Speeds

Consider the basic equations relating lift to air density and airspeed, as follows: Assuming a standard day,

$$W = 0.5C_{Ls}\left(\rho V_s^2\right)_{sd} S_w \qquad (6.3)$$

where C_{Ls} is the lift coefficient at stall, ρ is the air density for a standard day (slug/ft^3), V_s is the airspeed at stall for a standard day (ft/s), and S_w is the wing reference area (ft^2).

A similar equation can now be written for a hot day:

$$W = 0.5C_{Ls}\left(\rho V_s^2\right)_{hd} S_w \qquad (6.4)$$

Equating Eqs. (6.3) and (6.4), the relationship between air density and airspeed for a standard and hot day becomes

$$\left(\rho V_s^2\right)_{hd} = \left(\rho V_s^2\right)_{sd} \qquad (6.5)$$

If lift as defined in Eq. (6.3) is to be maintained constant, then as the air density is reduced due to a hot day with respect to the standard day reference, then the airspeed must be increased as noted in Eq. (6.5).

The term V_{L2} as defined by FAR 25.479(a)(2) may now be obtained from Eq. (6.5):

$$(V_s)_{hd} = (\rho_{sd}/\rho_{hd})^{\frac{1}{2}}(V_s)_{sd} \qquad (6.6)$$

From the perfect gas law:

$$\rho = P/(RT) \qquad \text{or} \qquad \rho T = \text{const} \qquad (6.7)$$

It follows then at a constant altitude that the relationship between air density and temperature may be stated as

$$(\rho T)_{hd} = (\rho T)_{sd} \qquad (6.8)$$

The stall speeds for a hot day may now be calculated from Eqs. (6.6) and (6.8):

$$V_{L2} = V_{sd}(T_{HD}/T_0)^{\frac{1}{2}} \qquad (6.9)$$

Table 6.1 Landing speed calculations, $S_w = 1951$ ft^2

Airplane data	Condition parameters			
Landing flap	30	30	25	25
Alt., ft	0	13,500	0	13,500
Max. landing weight, lb	198,000	198,000	198,000	198,000
V_{so}, keas	103.8	103.8	106.7	106.7
C_L stall	2.779	2.779	2.630	2.630
Tail wind, ktas	15	15	15	15
V_{L1}, ktas[a]	103.8	127.7	106.7	131.3
V_{L2}, ktas[b]	107.8	133.2	110.8	136.9
Min. landing speed, ktas	104	128	107	131
Max. landing speed, ktas	150	181	154	186
Design landing speeds selected:				
Landing flap		30		25
Max. landing weight, lb		198,000		198,000
Min. landing speed, ktas		104		107
Alt., ft		0		0
Max. landing speed, ktas[c]		181		186
Alt., ft		13,500		13,500

[a]$V_{L1} = V_{so}$ as shown in Eq. (6.1).
[b]V_{L2} is calculated using Eq. (6.10).
[c]Calculated using Eq. (6.2).

where V_{sd} is the stall speed for a standard day at the altitude under consideration. By defining the stall speed V_{sd} in terms of equivalent airspeed, Eq. (6.9) may be written as

$$V_{L2} = V_{so}\sigma^{-\frac{1}{2}}(T_{hd}/T_0)^{\frac{1}{2}} \qquad \text{(ktas)} \qquad (6.10)$$

where V_{so} is the stall speed at sea level for standard day (keas), T_{hd} is the hot day temperature (°R), T_0 is the standard day temperature (°R), and σ is the air density ratio, ρ/ρ_0.

Landing speeds calculated per FAR 25.479 are shown in Table 6.1. The hot day atmospheric parameters shown in Table 6.2 are obtained from Ref. 1.

The design landing speeds are selected as noted, whereby the minimum airspeed is obtained for the sea level condition and the maximum airpseed for landing is obtained at the maximum altitude for which the airplane is to be certified using Eq. (6.2). In this example the airplane is certified for a 15-kn tail wind; therefore per FAR 25.479(a)(3), the tail wind must be included in the calculation of the maximum landing speed.

6.3 Two-Point Landing Conditions

A two-point landing is defined as a landing whereby the airplane nose gear is held from contact with the ground, as shown in Figs. 6.1 and 6.2, until complete absorption of the energy of descent is accomplished.

Table 6.2 Hot day atmospheric parameters[a]

Alt., ft	$\sigma^{-0.5}$	T_0, °R	T_{hd}, °R	$(T_{hd}/T_0)^{0.5}$
0	1.0	518.7	559.7	1.039
10,000	1.1637	483.0	524.0	1.042
13,500	1.2303	470.6	511.6	1.043

where

T_0 = standard day temperature
$T_{hd} = T_0 + 41$, °R per FAR 25.479
$\sigma = (\rho/\rho_0)$ = density ratio
ρ = density of air at altitude
ρ_0 = density of air at sea level

[a]See Ref. 1.

Two attitudes must be considered: 1) level landing per FAR/JAR 25.479 with the nose gear just clear of the ground at airspeeds from $V_{landing\ sd}$ to $V_{landing\ hd}$ and 2) tail-down landing per FAR/JAR 25.481 as limited by structure (or tail bumpers) at airspeeds from $V_{landing\ sd}$ to $V_{landing\ hd}$, where the landing speeds are defined by Eqs. (6.1) and (6.2), and the descent velocities are as defined in Sec. 6.1.

6.3.1 Two-Point Level Landing Analysis

Referring to Fig. 6.1, the forces and moments acting on the airplane during a two-point level landing may be determined,

$$\Sigma F_z = 0: \qquad V_{MGr} + V_{MGl} = n_z W - L \tag{6.11}$$

or rearranging terms,

$$n_z W - L = V_{MGr} + V_{MGl} \tag{6.12}$$

Since $L = W$ by definition in FAR 25.473(a)(2),

$$W(n_z - 1) = V_{MGr} + V_{MGl} \tag{6.13}$$

The airplane incremental vertical load factor at the center of gravity may now be determined from Eq. (6.14):

$$\Delta n_z = (V_{MGr} + V_{MGl})/W \tag{6.14}$$

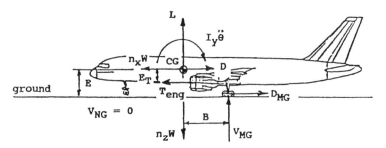

Fig. 6.1 Two-point level landing condition.

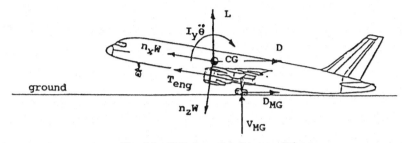

Fig. 6.2   Tail-down landing condition.

In a similar manner, the summation of forces in the fore and aft directions will give the longitudinal load factor at the airplane center of gravity:

$$\Sigma F_x = 0: \qquad D_{MGr} + D_{MGl} = n_x W - D + T_{eng} \qquad (6.15)$$

$$n_x = (D_{MGr} + D_{MGl} + D - T_{eng})/W \qquad (6.16)$$

If the assumption is made that engine thrust is equal to airplane drag during the landing, then longitudinal load factor may be determined as shown in Eq. (6.17):

$$n_x = (D_{MGr} + D_{MGl})/W \qquad (6.17)$$

By taking the summation of moments about the airplane center of gravity, the pitching acceleration required to maintain balance during the landing may be obtained:

$$\Sigma M_{cg} = 0: \qquad I y \ddot{\theta} = B(V_{MGr} + V_{MGl}) + E_{ax}(D_{MGr} + D_{MGl}) - E_T T_{eng} \qquad (6.18)$$

$$\ddot{\theta} = [B(V_{MGr} + V_{MGl}) + E_{ax}(D_{MGr} + D_{MGl}) - E_T T_{eng}]/I_y \qquad (6.19)$$

where $E_{ax} = E - rr$, rr is the rolling radius of the wheels (in.), E is the distance from airplane center of gravity to ground plane (in.), T_{eng} is the total engine thrust (lb), and D is the airplane drag in the landing configuration (lb).

The equations represented by Eqs. (6.11–6.19) as shown are applicable for a flexible or rigid airplane analysis. The resulting gear loads, load factors, and pitching acceleration will vary with time during the landing impact. By neglecting the engine thrust term in Eq. (6.19), the resulting pitching acceleration will be conservative. It should be noted that the aerodynamic pitching moment about the airplane center of gravity is assumed zero throughout the airplane oleo stroke.

6.3.2   Rigid Airplane Two-Point Level Landing Analysis

Assuming a rigid airplane but including the oleo and tire characteristics, the vertical load factor may be calculated for the two-point level landing condition at the time of maximum vertical ground reaction using Eq. (6.14):

$$n_z = 1 + (V_{MGr} + V_{MGl})/W \qquad (6.20)$$

where V_{MGr} and V_{MGl} have been obtained from drop test data or other acceptable analytical analyses.

Making the assumption that the drag load acting on the main landing gear is equal to 0.25 of the maximum vertical ground reaction per FAR 25.479(c)(2), one can compute the forward acting load factor from Eq. (6.17):

$$n_x = 0.25(V_{MGr} + V_{MGl})/W \tag{6.21}$$

Substitution of Eq. (6.20) into Eq. (6.21) will yield the relationship between the maximum vertical load factor obtained during the landing and the longitudinal load factor at that time:

$$n_x = 0.25(n_z - 1) \tag{6.22}$$

The pitching acceleration at the time of maximum vertical gear loads may be calculated from Eq. (6.19). By neglecting the thrust term, and assuming the specified relationship between the drag and vertical loads of 0.25, the equation for pitching acceleration becomes

$$\ddot{\theta} = (V_{MGr} + V_{MGl})(B + 0.25E_{ax})/I_y \tag{6.23}$$

6.3.3 Time of Maximum Vertical Ground Reaction

Using the equations developed for the rigid airplane two-wheel level landing analysis in Sec. 6.3.2, example load factors at the time of the maximum vertical reaction are shown in Table 6.3 for several jet transport aircraft.

It should be noted that the large variations in load factors reflect the oleo length and hydraulic characteristics designed into the main landing gears. Airplanes A and F have long stroke gears installed, hence the reduction in maximum load factors during landing.

6.3.4 Maximum Spin-Up and Spring-Back Conditions

During the contact of the wheels with the runway surface, two conditions specified in FAR 25.479 need to be considered.

1) Per FAR 25.479(c)(1), the condition of maximum wheel spin-up load, drag components simulating the forces required to accelerate the wheel rolling assembly

Table 6.3   Two-point level landing load factors (design landing weights are shown, and limit descent velocity = 10 fps)

Airplane	Wheels per main gear	Max. landing weight, lb	V_{MG},[a] lb	n_z[b]
A	4	247,000	137,800	2.12
B	2	135,000	93,200	2.38
C	2	161,000	107,600	2.34
D	2	114,000	96,900	2.70
E	2	121,000	97,000	2.60
F	4	198,000	120,000	2.21

[a]Maximum vertical ground reaction used for analysis.
[b]$n_z = 1 + 2V_{MG}/MLW$ [see Eq. (6.20)].

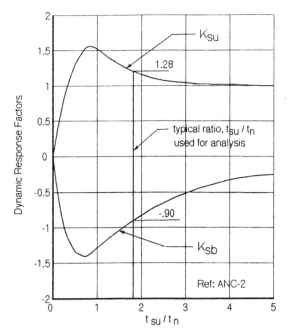

Fig. 6.3 Dynamic response factors for landing gear drag loads. t_n = natural period of landing gear in fore and aft mode, s; t_{su} = time required for wheel velocity to reach ground velocity, s; K_{su} = dynamic response factor for spin-up load; K_{sb} = dynamic response factor for spring-back load.

up to the specified ground speed, must be combined with the vertical ground reactions existing at the instant of peak drag load. The coefficient of friction need not exceed 0.80.

2) Per FAR 25.479(c)(3), the condition of maximum spring-back load, forward-acting horizontal loads resulting from a rapid reduction of the spin-up drag loads, must be combined with the vertical ground reactions existing at the instant of peak forward load.

The drag ratios for spin-up and spring-back landing conditions may be obtained from Ref. 2, which specifies the relationship between drag and vertical loads acting on the landing gear using a coefficient of friction of 0.55 times a dynamic response factor:

$$(D_{MG}/V_{MG}) = 0.55 K_{dyn} \tag{6.24}$$

The dynamic response factors K_{dyn} are shown in Fig. 6.3 as a function of the ratio of the time required for the wheels to obtain ground speed to the natural period of the landing gear in the fore and aft vibration mode. Examples of the response factors used for many large commercial jet aircraft are shown in this figure. These factors were verified (as conservative) during drop test of the gears or during the flight-test programs.

The ground contact coefficient of friction during the landing of 0.55 was obtained from Ref. 2 and is usually accepted by the certifying agencies.

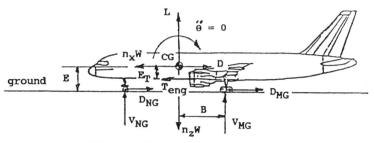

Fig. 6.4   Three-point level landing condition.

6.4   Three-Point Landing Conditions

A three-point landing is defined as a landing whereby the nose and main gears contact the runway simultaneously as shown in Fig. 6.4.

The three-point landing condition has a stipulation stated in FAR/JAR 25.479(e) (2) concerning the specified descent and forward velocities of the airplane, namely, "if reasonably attainable."

Depending on the airplane configuration, some rational landing analyses may not be possible within the design landing speeds defined by Eqs. (6.1) and (6.2). For these analyses some adjustment to the analysis may be required to determine nose gear loads during the landing, such as using lower landing flap settings or conservative landing speeds (higher than required by the regulations).

The three-point landing conditions are usually critical for the nose gear and related support structure, and the main gear landing loads are critical for the two-point landing conditions.

6.4.1   Three-Point Level Landing Analysis

The equations for a three-point landing may be determined in a similar manner as for the two-point landing. These equations apply as shown for a flexible or rigid airplane analysis.

Referring to Fig. 6.4,

$$\Sigma F_z = 0: \qquad V_{MGr} + V_{MGl} + V_{NG} = n_z W - L \qquad (6.25)$$

Since $L = W$ by definition in FAR 25.473(a)(2),

$$V_{MGr} + V_{MGl} + V_{NG} = W(n_z - 1) \qquad (6.26)$$

Rearranging terms as in Eq. (6.26), the vertical load factor acting at the airplane center of gravity for a three-point level landing may be determined:

$$n_z = 1 + (V_{MGr} + V_{MGl} + V_{NG})/W \qquad (6.27)$$

In a similar manner, the summation of forces in the fore and aft direction will give the longitudinal load factor at the airplane center of gravity:

$$\Sigma F_x = 0: \qquad D_{MGr} + D_{MGl} + D_{NG} = n_x W - D + T_{eng} \qquad (6.28)$$

$$n_x = (D_{MGr} + D_{MGl} + D_{NG} + D - T_{eng})/W \qquad (6.29)$$

For the three-point landing condition the assumption is made that the pitching moment is resisted by the nose gear, and hence the pitching acceleration is zero. This will maximize the nose gear load for the landing analysis:

$$\Sigma M cg = 0: \quad V_{NG}C - D_{NG}E_{NGa} = B(V_{MGr} + V_{MGl})$$
$$+ E_{MGa}(D_{MGr} + D_{MGl}) - E_T T_{eng} \tag{6.30}$$

where

$$E_{NGa} = E - rr_{NG}$$
$$E_{MGa} = E - rr_{MG}$$

and where rr_{MG} and rr_{NG} are the rolling radius of the main gear wheels and the nose gear wheels, respectively (in.), E is the vertical distance from airplane center of gravity to ground plane (in.), E_T is the vertical distance from airplane center of gravity to engine thrust line (in.), T_{eng} is the total engine thrust for the landing condition usually assumed $T_{eng} = D$ (lb), and D is the airplane drag in the landing configuration (lb).

6.4.2 Rigid Airplane Three-Point Level Landing Analysis

A rigid airplane three-point level landing analysis can be developed to solve for the nose gear load using the following assumptions.

1) During the landing $T_{eng} =$ drag.

2) The pitching moment due to engine thrust, if conservative to do so, may be neglected.

3) The relationship of the drag load at the time of maximum vertical load is 0.25 and occurs at the time of maximum vertical load on the gears.

Using these assumptions one can derive the equations for airplane longitudinal load factor and nose gear load:

$$n_x = 0.25(V_{MGr} + V_{MGl} + V_{NG})/W \tag{6.31}$$
$$V_{NG} = (V_{MGr} + V_{MGl})(B + 0.25E_{MGa})/(C - 0.25E_{NGa}) \tag{6.32}$$

Combining Eqs. (6.27) and (6.31), the equation for longitudinal load factor is the same as Eq. (6.22) for the two-wheel level landing condition:

$$n_x = 0.25(n_z - 1) \tag{6.22}$$

Combining Eqs. (6.27) and (6.32), one can derive the nose gear load during a three-point level landing:

$$V_{NG} = (n_z - 1)WF/(1 + F) \tag{6.33}$$

where

$$F = (B + 0.25E_{MGa})/(C - 0.25E_{NGa}) \tag{6.34}$$

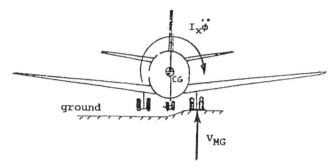

Fig. 6.5   One-gear landing condition.

6.5   One-Gear Landing Conditions

For the one-gear landing conditions, the airplane is assumed to be in the level attitude and to contact the ground on one main landing gear as shown in Fig. 6.5, in accordance with FAR 25.483. In this attitude, 1) the ground reactions must be the same as those obtained on that side per 25.479(c)(2), which defines the maximum vertical load condition for the two-wheel level landing condition (see Sec. 6.3.1), and 2) each unbalanced external load must be reacted by airplane inertia in a rational or conservative manner.

6.6   Side Load Conditions

For the side load condition, the airplane is assumed to be in the level attitude with only the main wheels in contact with the runway as shown in Fig. 6.6 per FAR 25.485.

Side loads of 0.8 of the vertical reaction (on one side) acting inward and 0.6 of the vertical reaction (on the other side) acting outward must be combined with one-half the maximum vertical ground reactions obtained in the level landing conditions. These loads are assumed to be applied at the ground contact point and to be resisted by the inertia of the airplane. The drag loads may be assumed zero.

6.7   Rebound Landing Conditions

The rebound criteria as stipulated in FAR 25.487 provide the design requirements for the landing gear and its supporting structure for the loads occurring during rebound of the airplane from the landing surface.

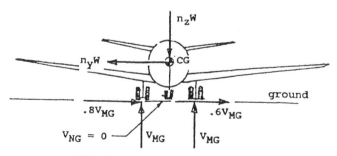

Fig. 6.6   Side load landing condition.

Table 6.4 Landing gear drop test requirements

	Gross weight	Descent velocity, ft/s	Drop height, in.
Design conditions	Max. landing weight	10	18.7
	Max. takeoff weight	6	6.7
Reserve energy condition	Max. landing weight	12	27.0

With the landing gear fully extended and not in contact with the ground, a load factor of 20.0 (limit) must act on the unsprung weights of the landing gears. This load factor must act in the direction of motion of the unsprung weights as they reach their limiting extended positions in relation to the sprung parts of the landing gear.

6.8 Landing Gear Shock Absorption and Drop Tests

Landing gear shock absorption and drop tests must be made in accordance with FAR 25.723, 25.725, and 25.727 for takeoff and landing weights as summarized in Table 6.4.

The drop heights may be calculated by relating the kinetic energy required for the landing condition to the potential energy for the drop test:

$$KE = \tfrac{1}{2}(W/g)v^2 \qquad \text{(in.-lb)} \qquad (6.35)$$

$$PE = Wh \qquad \text{(in.-lb)} \qquad (6.36)$$

Equating Eqs. (6.35) and (6.36), one can determine the drop height as a function of the landing descent velocity:

$$\text{drop height} = h = \tfrac{1}{2}v^2/g \qquad (6.37)$$

where v is the landing descent velocity (ft/s), and g is the acceleration of gravity (ft/s^2).

Energy absorption tests are required per FAR 25.723 to show that the limit design load factors will not be exceeded for takeoff and landing weights. Analyses based on previous tests conducted on the same basic landing gear system may be used for increases in previously approved takeoff and landing weights.

Reserve energy shock absorption tests must be accomplished simulating a descent velocity of 12 ft/s at design landing weight per FAR 25.723(b).

Examples of drop test data for main and nose gears are shown in Figs. 6.7 and 6.8, respectively.

Effective weights to be used in drop tests are defined in FAR 25.725. The calculations of effective weights for a main gear drop test are shown in Table 6.5 for a cargo airplane with a lateral unbalance. The calculations of effective weights for a nose gear drop test are shown in Table 6.6.

6.9 Elastic Airplane Analysis

According to the requirements of FAR/JAR 25.473, the method of analysis must also include significant structural dynamic response during the landing. For jet aircraft transports certified before 1968, what has been called a "dynamic landing

Table 6.5 Main gear effective weight calculations for two-point level landing (the airplane shown in this table has a lateral unbalance due to cargo landing)

Weight cond.	W_L, lb	BL_{cg}, in.	W_{EMGr},[a] lb	W_{EMGl},[b] lb
Max. landing weight	103,000	4.45	53,720	49,280
Max. takeoff weight	114,000	4.70	59,510	54,490

[a] $W_{EMGr} = W_L(0.50 + BL_{cg}/T)$.

[b] $W_{EMGl} = W_L - W_{EMGr}$, where W_L is the landing weight (lb), T is the lateral distance between main landing gears and $= 206.0$ in. (for this example), and BL_{cg} is the lateral unbalance (in.).

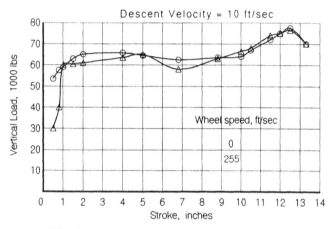

Fig. 6.7   Example of main gear drop test results.

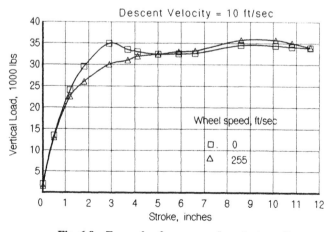

Fig. 6.8   Example of nose gear drop test results.

Table 6.6   Nose gear effective weight calculations for three-point level landing per FAR 25.725(b)

Weight cond.	W_L, lb	CG, % mac/100	B, in.	E, in.	W_{ENG},[a] lb
Max. landing weight	103,000	0.05	65.9	120.3	21,075
Max. takeoff weight	114,000	0.05	65.9	118.6	23,218

[a] $W_{ENG} = W_L [B + 0.25(E - R_{rr})]/C$. For this example, $C = 448.9$ in., and R_{rr} is the rolling radius and $= 16.5$ in.

analysis" was accomplished using analog computers that limited the number of degrees of freedom in the analyses. The intent of these analyses was to determine the landing gear loads and the loads (or load factors) acting on large mass items such as external fuel tanks and nacelles.

With the advent of digital computer capabilities, many degrees of freedom may now be incorporated to represent structural dynamic characteristics of the wing, fuselage, landing gear, support structure flexibility, and large mass items such as engines supported either in nacelles on the wings or body mounted like the 727, DC-10, and MD-80 series type airplanes. The response load factors acting on aft body-mounted auxiliary power units (APUs) may also be included.

Because of the complexity of dynamic landing analyses, readers are encouraged to review other sources for discussions on the equations of motion, gear oleo pneumatic characteristics, and the airplane elastic representation.

6.10   Automatic Ground Spoilers

Some aircraft configurations are designed such that the ground spoilers on the wing are automatically applied when the gear comes in contact with the ground. The intent is to dump airload from the wing to increase brake effectiveness, thus decreasing the landing roll-out.

Depending on the design characteristics of the system, the assumption that lift is maintained equal to airplane weight during the initial stroke of the gears becomes questionable. If the spoilers are applied too soon, then the analysis must include this effect, which will increase the design landing loads on the aircraft structure. Generally a time delay is incorporated in the activation system to extend spoilers after the gear has completed the initial stroke during landing.

References

[1] Anon., "Manual of the ICAO Standard Atmosphere, Calculations by the NACA," NACA TN 3182, May 1954.

[2] Anon., "Ground Loads," ANC-2 Bulletin, U.S. Depts. of the Air Force, Navy, and Commerce, Civil Aeronautics Administration, Oct. 1952.

7
Ground-Handling Loads

Ground-handling loads, although not complex in nature (except for multiple gear aircraft such as the 747 and DC-10), pose some interesting problems, such as braking conditions and the special case of airplane tie down, which sometimes is called tethering.

From a historical perspective, ground load requirements have not changed since the design of the early 707/DC-8 aircraft except for design considerations for nose gear loads due to abrupt braking.

This criterion was not a part of FAR 25.493 but has been applied as a special condition by the British Civil Aviation Authority (CAA). In the harmonization process of 1993, a change to FAR/JAR 25.493 is proposed to include the dynamic reaction effect on nose gear loads as a result of sudden application of maximum braking force.

The tethering problem, although never included, is of importance in providing the capability for tie down of the airplane in very high winds. This is an airline problem and is of particular concern for operators in the Pacific Rim area.

7.1 Ground-Handling Conditions

The ground-handling loads as discussed in this chapter are a set of conditions involving ground maneuvers, braking during landing and takeoff, and special conditions for towing, jacking, and tethering. For static analysis conditions, airloads are assumed zero, and only inertia loads are calculated as required for an analysis equilibrium. Ground-handling conditions are usually defined as taxi conditions, braked-roll conditions, refused takeoff conditions, turning conditions, towing and jacking conditions, and the tethering problem.

Ground load analysis geometric parameters and load sign conventions are defined in Fig. 7.1. An example set of landing gear loads is shown in Table 7.1 for an aircraft that meets the requirements of FAR/JAR 25.493 for a rigid airplane.

7.2 Static Load Conditions

Static load conditions are defined at $n_z = 1.0$ with the airplane in a three-point static attitude as shown in Fig. 7.2. Assuming a standard three-post gear configuration as shown in this figure, the equations for nose and main gear loads may be derived.

Main gear loads:

$$(V_{MGs})_R = W(A/2C + BL_{cg}/T) \tag{7.1}$$

$$(V_{MGs})_L = W(A/2C - BL_{cg}/T) \tag{7.2}$$

$$(D_{MGs})_R = (D_{MGs})_L = 0 \tag{7.3}$$

$$(S_{MGs})_R = (S_{MGs})_L = 0 \tag{7.4}$$

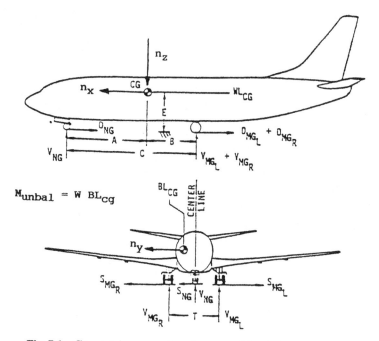

Fig. 7.1   Geometric parameters for ground-handling conditions.

Nose gear loads:

$$V_{NGs} = WB/C \tag{7.5}$$

$$D_{NGs} = S_{NGs} = 0 \tag{7.6}$$

where s is the static condition at $n_z = 1.0$.

7.3   Taxi, Takeoff, and Landing Roll Conditions

The ground-handling load requirements for taxi, takeoff, and landing roll conditions as proposed by the FAR/JAR 1993 harmonization process are grouped together as follows.

FAR/JAR 25.491: "Within the range of appropriate ground speeds and approved weights, the airplane structure and landing gear are assumed to be subjected to loads not less than those obtained when the aircraft is operated on the roughest ground that may reasonably be expected in normal operation."

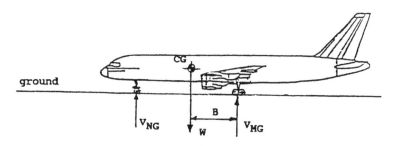

Fig. 7.2   Airplane static condition.

Table 7.1a Design ground loads for right main gear only of a cargo airplane with a lateral unbalance

Weight cond.	W, lb	CG, % mac/100	M_{unbal}, in.-lb	BL_{cg}, in.	A, in.	E, in.
MTW	120,000	0.25	500,000	4.17	411.8	110.0
MLW	105,000	0.34	500,000	4.76	423.9	110.0

$$C, \text{ in.} = 450.0 \qquad T, \text{ in.} = 210.0$$

Cond. type	n_z	n_y	Coefficient friction	V_{MGr}, lb ult.[a]	D_{MGr}, lb ult.[a]	S_{MGr} lb ult.[a]
Main gear loads at maximum taxi weight (MTW)						
Two-point braked roll	1.0	0	0.80	93,600	74,900	0
Three-point braked roll	1.0	0	0.80	72,500	58,000	0
Unsymmetrical braked roll	1.0	0	0.80	78,300	62,600	7,000
Reversed braked roll	1.0	0	0.55	85,900	−47,200	0
Ground turn	1.0	−0.50	0	133,100	0	−66,500
Taxi/takeoff	2.0	0	0	171,800	0	0
Pivot	1.0	0	0.80	85,900	0	0
Towing	1.0	0	0	85,900	27,000	0
Main gear loads at maximum landing weight (MLW)						
Two-point braked roll	1.2	0	0.80	98,800	79,000	0
Three-point braked roll	1.2	0	0.80	78,700	63,000	0

[a]$SF = 1.5$.

The preceding criterion has not changed significantly since the design of the 707 airplanes established in 1953.

7.3.1 Historical Perspective

The minimum load factor for takeoff was stipulated in Ref. 1 as 2.0 limit with the airplane at maximum takeoff gross weight. The airplane was to be in a three-point attitude with zero drag and side loads acting on the gears. This was the basis of the so-called 3.0-g (ultimate) requirement and was used for aircraft designed before 1953.

During the certification process of the 707-100 airplane, the advent of the so-called "bogey gear" (four-wheel truck on each main gear) was cause for concern in determining the design load factor to be used in computing taxi loads. It was felt that the four-wheel truck had the capability of "walking over bumps," hence attenuating the load factor. An analysis of the test data obtained from the B-36 airplane, which had a similar gear configuration, was undertaken to verify design load factors. This analysis subsequently allowed reduction of the design taxi factor to 2.50 ultimate (1.67 limit) for aircraft with this type of gear configuration.

Table 7.1b   Design ground loads for nose gear

Weight cond.	W, lb	CG, % mac/100	M_{unbal}, in.-lb	BL_{cg}, in.	B,[a] in.	E, in.
MTW	120,000	0.070	500,000	4.17	62.6	110.0
MLW	105,000	0.050	500,000	4.76	65.3	110.0

C, in. $= 450.0$ T, in. $= 210.0$

Cond. type	n_z	n_y	Coefficient friction	V_{NG}, lb ult.[b]	D_{NG}, lb ult.[b]	S_{NG}, lb ult.[b]
Nose gear loads at maximum taxi weight (MTW)						
Two-point braked roll	1.0	0	0.80	0	0	0
Three-point braked roll	1.0	0	0.80	50,400	0	0
Unsymmetrical braked roll	1.0	0	0.80	39,500	0	−9,700
Reversed braked roll	1.0	0	0.55	25,000	0	0
Ground turn	1.0	−0.50	0	25,000	0	−12,500
Taxi/takeoff	2.0	0	0	50,000	0	0
Nose-gear yaw	1.0	0	0.80	25,000	0	±20,000
Towing	1.0	0	0	25,000	27,000	0
Nose gear loads at maximum landing weight (MLW)						
Two-point braked roll	1.2	0	0.80	0	0	0
Three-point braked roll	1.2	0	0.80	53,800	0	0

[a] $B = C - A$.   [b] $SF = 1.5$.

7.3.2 Taxi Design Load Factors

The design load factors used for gear load calculations vary with the airplane configuration and time period in which the aircraft structure was designed. The load factors as used for rigid loads analyses of various aircraft are summarized in Table 7.2. The regulations do not specifically require that a given load factor be used for design, but rather the interpretation as noted in FAR/JAR 25.491.

As computer technology improved, the capability of performing a dynamic taxi analysis became possible, and later model aircraft were assessed considering structural dynamic effects and landing gear hydraulic characteristics. A profile of the San Francisco runway no. 28R, before refurbishment, as shown in Fig. 7.3 is considered acceptable for meeting the requirements of FAR/JAR 25.491.

7.3.3 Taxi Gear Loads

The equations for calculating gear loads for taxi, takeoff, and landing roll-out assuming a rigid airplane analysis are obtained from the static equations defined in Eqs. (7.1–7.6), but at the selected design taxi load factor.

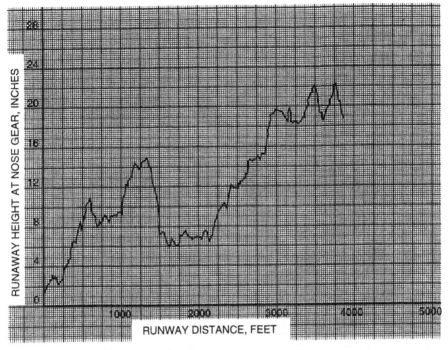

Fig. 7.3   San Francisco runway profile.

Nose gear loads:

$$V_{NG} = n_z V_{NGs} \tag{7.7}$$

$$D_{NG} = S_{NG} = 0 \tag{7.8}$$

Main gear loads:

$$(V_{MG})_R = n_z (V_{MGs})_R \tag{7.9}$$

$$(V_{MG})_L = n_z (V_{MGs})_L \tag{7.10}$$

$$(D_{MG})_R = (S_{MG})_R = 0 \tag{7.11}$$

$$(D_{MG})_L = (S_{MG})_L = 0 \tag{7.12}$$

Table 7.2   Design load factors for taxi,
takeoff, and landing roll-out for rigid analysis

Aircraft	n_z, limit	Main gear configuration
A1	1.67	Four-wheel truck
A2	2.0	Two wheels per gear
A3	2.0	Four-wheel truck
A4	1.67	Multiple gears

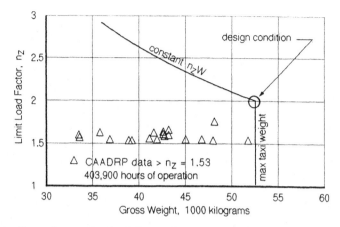

Fig. 7.4 Ground operation load factors for airplane with two-wheel main gears.

where s represents the static conditions defined by Eqs. (7.1–7.6), and n_z is the limit design load factor selected.

These equations apply to aircraft with two main gears and one nose gear and must be modified for multiple main gear configurations such as the 747 airplanes. The use of the term "rigid analysis" refers to the assumption that the airplane is treated as a rigid body vs the full dynamic analysis discussed in Sec. 7.3.2. An example of the gear loads using these equations are shown in Table 7.1.

7.3.4 Operational Experience

Vertical load factors obtained from operational experience for two aircraft from the British Airlines fleet as reduced from Civil Aircraft Airworthiness Data Recording Program (CAADRP) data are summarized in Figs. 7.4 and 7.5. These data were obtained from flights between April 1983 and September 1989. The resulting load factors are compared with limit design load factors for the ground-handling taxi, takeoff, and landing roll conditions. The statistical analysis of the data had a load factor cutoff of $\triangle n_z = +/- 0.50$.

7.4 Braked-Roll Conditions

The ground-handling loads for the braked-roll requirements of FAR/JAR 25.493 result in three basic conditions. The first two are the traditional braked-roll conditions applied to aircraft certified before 1980.

During certification by the British CAA of the current series of jet aircraft after 1981, consideration was given for a new design condition for nose gear and related structure. In February 1993 this condition was proposed for incorporation into FAR 25.493 as part of the harmonization process.

7.4.1 Three-Point Braked-Roll per FAR/JAR 25.493(b)(1)

The airplane is assumed to be in a level attitude and the loads distributed between the main gears and the nose gear (or tail wheel). Zero pitching acceleration is assumed, and no wing lift is considered.

The equations for determining gear loads for braking conditions in the three-point attitude may be derived from the forces acting on the airplane as shown in

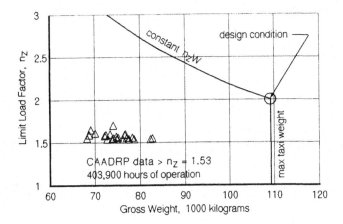

Fig. 7.5   Ground operation load factors for airplane with four-wheel truck main gears.

Fig. 7.6. The equations represented by (7.13–7.21) must be modified for airplane configurations with nose gear brakes.

Nose gear loads:

$$V_{NG} = n_z W (B + E\mu_{MG})/(C + E\mu_{MG}) \tag{7.13}$$

where assuming no nose gear brakes,

$$D_{NG} = 0 \tag{7.14}$$

$$S_{NG} = [D_{MGl}(BL_{\text{cg}} + 0.50T) + D_{MGr}(BL_{\text{cg}} - 0.50T)]/C \tag{7.15}$$

Main gear loads:

$$V_{MGr} = n_z W(0.50 + BL_{\text{cg}}/T) - 0.50V_{NG} \tag{7.16}$$

$$V_{MGl} = n_z W - V_{NG} - V_{MGr} \tag{7.17}$$

$$D_{MGr} = \mu_{MG} V_{MGr} \tag{7.18}$$

$$D_{MGl} = \mu_{MG} V_{MGl} \tag{7.19}$$

$$S_{MGr} = -0.50 S_{NG} \tag{7.20}$$

$$S_{MGl} = +0.50 S_{NG} \tag{7.21}$$

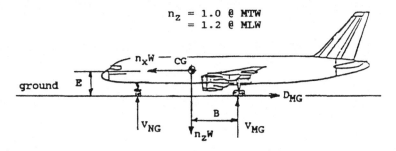

Fig. 7.6   Three-point braked-roll condition.

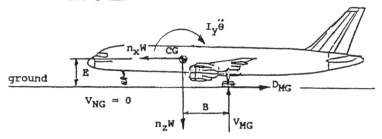

$$n_z = 1.0 \ @ \ MTW$$
$$= 1.2 \ @ \ MLW$$

Fig. 7.7 Two-point braked-roll condition.

where $\mu_{MG} = 0.80$ or as limited by brake torque (see Sec. 7.4.3), $n_z = 1.0$ limit at design taxi weight, and $n_z = 1.2$ limit at design landing weight.

Pitching acceleration about airplane center of gravity:

$$\ddot{\theta} = 0 \qquad (7.22)$$

7.4.2 Two-Point Braked-Roll per FAR/JAR 25.493(b)(2)

The airplane is assumed to be in a level attitude with the nose gear off the ground with the resulting pitching moment reacted by angular acceleration. No wing lift may be considered.

The equations for determining gear loads for braking conditions in the two-point attitude may be derived from the forces acting on the airplane shown in Fig. 7.7.

Nose gear loads:

$$V_{NG} = D_{NG} = S_{NG} = 0 \qquad (7.23)$$

Main gear loads:

$$V_{MGr} = n_z W (0.50 + B L_{cg}/T) \qquad (7.24)$$
$$V_{MGl} = n_z W - V_{MGr} \qquad (7.25)$$
$$S_{MGr} = S_{MGl} = 0 \qquad (7.26)$$
$$D_{MGr} = \mu_{MG} V_{MGr} \qquad (7.27)$$
$$D_{MGl} = \mu_{MG} V_{MGl} \qquad (7.28)$$

where $\mu_{MG} = 0.80$ or as limited by brake torque (see Sec. 7.4.3), $n_z = 1.0$ limit at design taxi weight, and $n_z = 1.2$ limit at design landing weight.

Pitching acceleration about airplane center of gravity:

$$\ddot{\theta} = -[B n_z W + E (D_{MGr} + D_{MGl})]/I_y \qquad (7.29)$$

7.4.3 Drag Reaction

A coefficient of friction of 0.80 is assumed for all conditions except where limited by maximum brake torque [see FAR/JAR 25.493(c)]. The drag reaction is applied at the ground contact point with gears with brakes as noted in Fig. 7.8.

direction of roll

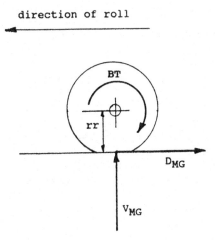

Fig. 7.8   Landing gear drag load due to brake torque.

Landing gear drag loads due to brake torque are calculated using Eq. (7.30):

$$D_{MG} = BT_{whl} n_{whl}/rr \qquad (7.30)$$

where BT_{whl} is the brake torque limit per wheel, (ft-lb), n_{whl} is the number of wheels per gear, and rr is the rolling radius (ft). (See Fig. 7.8.)

The resulting braking coefficient μ_{MG} may now be defined as

$$\mu_{MG} = D_{MG}/V_{MG} \qquad (7.31)$$

By taking moments about the main gear ground contact point, the equation for determining nose gear load for brake torque limited conditions may be derived:

$$V_{NG} = [n_z WB + E(D_{MGr} + D_{MGl})]/C \qquad (7.32)$$

By substituting Eqs. (7.17) and (7.31) into Eq. (7.32), one can derive Eq. (7.13).

7.4.4   Nose Gear Loads Due to Sudden Application of Brakes per FAR/JAR 25.493(d)

For airplanes with a nose gear, the nose gear and airplane must be designed to withstand the loads arising from an increase of vertical dynamic reaction at the nose gear as a result of sudden application of maximum braking force, taking into account the dynamic pitching moment of the airplane. This condition is at design takeoff weight with the nose and main gears in contact with the ground and with a steady vertical load factor of 1.0.

The steady-state nose gear reaction must be combined with the maximum incremental nose gear vertical reaction caused by the sudden application of maximum braking force. Equation (7.33), for determining the nose gear loads, is as given in the proposed regulations:

$$V_{NG} = n_z W[B + (f\mu_{MG}AE)/(C + \mu_{MG}E)]/C \qquad (7.33)$$

where $n_z = 1.0$ limit, W is the maximum taxi weight (lb), and $\mu_{MG} = 0.80$, and where the dynamic response factor $f = 2.0$ unless a lower factor is substantiated using Eq. (7.34),

$$f = \text{dynamic factor} = 1 + e^b \tag{7.34}$$

$$b = (-\pi\tau)/(1 - \tau^2)^{\frac{1}{2}} \tag{7.35}$$

where τ is the effective critical damping ratio of the rigid-body pitching mode about the main gear effective ground point.

A coefficient of friction of 0.80 is assumed.

In the absence of a rational analysis, the nose gear load must be calculated per Eq. (7.33). It should be pointed out that this condition applies only to the calculation of nose gear loads and is not to be applied to the calculation of design drag loads for the main gears.

Nose gear loads calculated using Eq. (7.33) are shown in Table 7.3. Three conditions are shown using a dynamic response factor of 2.0, whereas the fourth condition is based on a dynamic response factor calculated using Eq. (7.34) for airplane C. All three of the aircraft shown have the same nose gear and support structure but different lengths between the main gears and nose gear.

7.4.5 Reversed Braking per FAR/JAR 25.507

Gear loads due to reversed braking are calculated assuming a drag load applied in the forward direction equal to 0.55 of the vertical load at each wheel with brakes. This drag load need not exceed the load developed by 1.2 times the nominal maximum static brake torque.

Table 7.3 Nose gear loads due to sudden application of brakes per FAR/JAR 25.493(d); main gear braking coefficient, $\mu_{MG} = 0.80$

	Airplane[a]			
	A	B	C	C
Weight W, lb	135,500	139,800	134,000	134,000
CG, % mac/100	0.13	0.05	0.12	0.12
n_z limit	1.0	1.0	1.0	1.0
A, in.	435.30	496.63	380.03	380.03
B, in.	54.50	65.67	56.28	56.28
C, in.	489.80	562.30	436.30	436.30
E, in.	100.00	102.60	100.50	100.50
Dynamic factor f	2.0	2.0	2.0	1.81[b]
V_{NG} limit	48,890	47,780	53,610	50,160

[a]All three aircraft shown have the same nose gear and supporting structure.
[b]Based on Eq. (7.34) using an effective critical damping ratio of the rigid-body pitching mode of 0.064 ($g = 0.128$).

The airplane must be in a three-point static ground attitude at $n_z = 1.0$. For airplanes with nose wheels, the pitching moment is balanced by rotational inertia.

Nose gear loads:

$$V_{NG} = n_z V_{NGs} \tag{7.36}$$

$$D_{NG} = S_{NG} = 0 \tag{7.37}$$

Main gear loads:

$$V_{MGr} = n_z V_{MGrs} \tag{7.38}$$

$$V_{MGl} = n_z V_{MGls} \tag{7.39}$$

$$D_{MGr} = -\mu_{MG} V_{MGr} \tag{7.40}$$

$$D_{MGl} = -\mu_{MG} V_{MGl} \tag{7.41}$$

$$S_{MGr} = S_{MGl} = 0 \tag{7.42}$$

where $n_z = 1.0$ limit, $\mu_{MG} = 0.55$, and s is the static conditions defined by Eqs. (7.1–7.6).

Pitching acceleration about airplane center of gravity:

$$\ddot{\theta} = -[E(D_{MGr} + D_{MGl})]/I_Y \tag{7.43}$$

7.5 Refused Takeoff Considerations

As a supplement to the braked-roll requirements of FAR/JAR 25.493, consideration will be given to the gear loads developed during refused takeoff (RTO) flight testing of the airplane for certification. Many times these tests are accomplished at increased gross weights to cover near-future growth versions of the airplane. An understanding of the gear loads applied to the aircraft structure during these tests is necessary to allow such testing to be accomplished.

FAR/JAR 25.489 states that for ground-handling conditions no wing lift may be considered. For the refused takeoff conditions the activation of spoilers on the wing to increase braking effectiveness may result in a negative lift depending on the flap setting used. The effect on gear loads is as if the airplane gross weight was increased.

During takeoff as the aircraft approaches V_1 speed, the takeoff decision speed, the takeoff may safely be aborted. Beyond this speed the pilot can get the airplane airborne. This testing is accomplished during the certification program to demonstrate takeoff field length requirements. After V_1 is obtained, engine thrust is cut, ground spoilers are extended (if available), and maximum braking effort is applied.

7.5.1 Main and Nose Gear Loads

The equations for determining gear loads during an RTO are similar to the three-point braked-roll conditions shown in Sec. 7.4.1, except that airplane lift is included. As will be shown by the example in Table 7.4, lift coefficients in the takeoff attitude with ground spoilers extended are negative.

Defining the lift at the time of application of brakes during a refused takeoff as

$$L = C_{Lrto} V_e^2 S_w / 295 \tag{7.44}$$

Table 7.4 Gear loads for RTO analysis assuming the following constants: $C = 720.3$ in., $T = 288$ in., $BL_{cg} = 0$, and $n_z = 1.0$ [the data shown are used to plot the curves in Figs. 7.9 and 7.10, a conservative braking torque is assumed, see Eqs. (7.45) and (7.46)]

Flap position	W, lb	CG, % mac/100	D_{MG},[a] lb/gear	V_{MG},[a] lb/gear	V_{NG},[a] lb
1	250,000	0.12	0	144,974	44,265
			75,000[b]	128,907	76,397
5	250,000	0.12	0	128,205	39,145
			75,000[b]	112,138	71,277
15	250,000	0.12	0	118,094	36,057
			75,000[b]	102,027	68,190
1	250,000	0.25	0	150,982	32,247
			75,000[b]	134,916	64,380

[a] All loads shown are limit ($SF = 1.0$).
[b] Maximum brake torque limit above speeds of 100 kn is assumed for this analysis.
Note: V_1 speeds at the preceding gross weights as used for this analysis are as follows:

Flap position	V_1 keas	lift, lb
1	180	−84,212
5	165	−45,554
15	155	−22,245

where C_{Lrto} is the lift coefficient at the time of application of brakes, including the effect of takeoff flaps and ground spoilers.

Modifying Eqs. (7.13), (7.16), and (7.17) to include lift, one can derive the equations for the RTO analysis:

$$V_{NG} = [(n_z W - L)B + E(D_{MGr} + D_{MGl})]/C \qquad (7.45)$$

$$V_{MGr} = (n_z W - L)(0.50 - BL_{cg}/T) - 0.50V_{NG} \qquad (7.46)$$

$$V_{MGl} = (n_z W - L) - V_{MGr} - V_{NG} \qquad (7.47)$$

At V_1 speeds the main gear drag loads are usually limited by brake torque and may be calculated using Eq. (7.30).

7.5.2 Solution to RTO Loads Analysis

Main and nose gear loads may be calculated directly from Eqs. (7.45–7.47) knowing the main gear drag loads as limited by brake torque. These loads then are compared with the maximum loads obtained from the design conditions discussed previously in this chapter.

RTO analysis loads are shown in graphical form in Fig. 7.9 for the main gear and Fig. 7.10 for the nose gear for an example airplane. The V_1 speeds shown are a function of the takeoff flap settings used and are shown in the example calculations

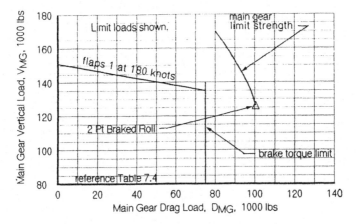

Fig. 7.9 Main gear loads for example RTO analysis.

in Table 7.4. Assuming a conservative maximum brake torque capability for speeds greater than 100 kn, the resulting main gear loads are less than the limit design conditions usually obtained from the two-point braked-roll condition.

As noted in Fig. 7.10, the nose gear loads for the maximum brake torque limit assumed for this example are quite high, exceeding the three-point braked-roll condition calculated from Eq. 7.13. The analysis data shown in this figure are conservative as the brake torque limit used cannot be obtained at the V_1 speeds shown in this example due to the design of the antiskid braking system.

A complete dynamic analysis could be accomplished assuming a ramp time history of the application of brakes using maximum brake torque actually available at the V_1 speeds for the configuration being investigated (lower than the value used for the example) to remove any conservatism in this type of analysis.

Suffice to say that the method as presented in this chapter allows for easy assessment of the structural capability for overweight operation during refused takeoff (RTO) certification tests.

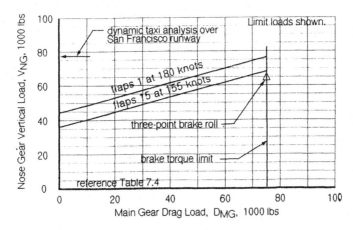

Fig. 7.10 Nose gear loads for RTO analysis.

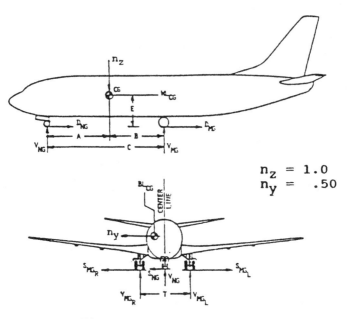

$$n_z = 1.0$$
$$n_y = .50$$

Fig. 7.11   Ground turn condition.

7.6   Turning Conditions

The ground-handling conditions involving turning conditions discussed in this book are ground turning, a steady turn executed by nose gear steering or differential thrust (shown in Fig. 7.11); nose-wheel yaw, a condition where the turn causes the nose gear to skid, thus producing a side load on the nose gear; unsymmetrical braking, application of brakes on one side of the airplane; nose gear steering, application of full normal steering torque; and pivoting conditions, pivoting about one side of the airplane with the brakes on that side locked.

7.6.1   Ground Turning per FAR/JAR 25.495

In the static position, the airplane is assumed to execute a steady turn by nose gear steering or by application of sufficient differential power, so that the limit load factors applied at the airplane center of gravity are

$$n_z = 1.0 \tag{7.48}$$
$$n_y = 0.50 \tag{7.49}$$

The side ground reaction of each wheel must be 0.50 of the vertical reaction.

The equations for ground turn gear loads are determined directly from the static load equations defined in Sec. 7.2.

Nose gear loads:

$$V_{NG} = n_z(V_{NG})_s \tag{7.50}$$
$$D_{NG} = 0 \tag{7.51}$$
$$S_{NG} = -n_y(V_{NG})_s \tag{7.52}$$

Main gear loads:

$$V_{MGr} = n_z(V_{MGr})_s + n_y WE/T \qquad (7.53)$$

$$V_{MGl} = n_z(V_{MGl})_s - n_y WE/T \qquad (7.54)$$

$$D_{MGr} = D_{MGl} = 0 \qquad (7.55)$$

$$S_{MGr} = -0.50 V_{MGr} \qquad (7.56)$$

$$S_{MGl} = 0.50 V_{MGl} \qquad (7.57)$$

where s is the static condition defined by Eqs. (7.1), (7.2), and (7.5).

The equations shown are for a left-hand ground turn. For a right-hand ground turn the signs are reversed in the second term of Eqs. (7.53) and (7.54) and in Eqs. (7.56) and (7.57).

7.6.2 Flight Test Limitations for Ground Turn Conditions

During the flight test program, overweight testing may be required as anticipated gross weight increases. These tests may include refused takeoff performance ground tests or flight tests such as stalls or handling characteristics.

Figure 7.12 relates ground speed to the turning radius for the design maximum taxi weight at the design side limit load factor of 0.50. An overweight operation at a reduced side load factor is shown that will give the same main gear load as the design gear load condition. By controlling ground speed and turn radius, the ground tests may be accomplished without exceeding the limit structural capability of the airplane. Placards for this type of operation are valid only while the airplane is operating with an experimental flight test certificate and may not be used in commercial service for overweight operations.

The relationship between turn radius, side load factor, and ground speed is seen in Eq. (7.58):

$$\text{Radius} = (GS)^2/(gn_y) \qquad \text{(ft)} \qquad (7.58)$$

where $GS = 1.6896$ (ground speed in knots) (ft/s), n_y is the lateral load factor developed during the turn, and $g = 32.2$ ft/s^2.

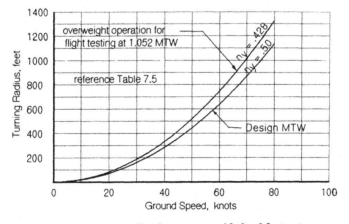

Fig. 7.12 Turning radius for constant side load factor turn.

Table 7.5 Turn radius for ground turn conditions calculated using Eq. (7.58) assuming the following constants: $c = 562.3$ in., $T = 206$ in., and n_y is the lateral load factor developed during the turn

	n_y, limit	W, lb	CG, % mac/100	A, in.	E, in.	V_{MGs},[a] lb limit	V_{MG},[b] lb limit
Design ground turn condition	0.50	143,000	0.28	527.55	102.2	67,081	102,554
Overweight operation	0.428	150,500	0.28	527.55	102.2	70,600	102,554

	Ground speed, kn						
	20	30	40	50	60	70	80
Turn radius, ft							
at $n_y = 0.50$	71	160	284	443	638	869	1135
at $n_y = 0.428$	83	186	331	518	746	1015	1326

[a]See Eq. (7.1).
[b]See Eq. (7.53).

Calculation of the turning radius for ground turn conditions is shown in Table 7.5 for the design side load factor and a reduced factor for flight testing at increased gross weights.

7.6.3 Nose–Wheel Yaw per FAR/JAR 25.499(a)

Nose–wheel yaw is caused by a ground turn such that the nose gear wheels skid, thus producing a side load at the nose gear wheel ground contact equal to 0.80 of the vertical ground reaction at this point. A vertical load factor of 1.0 at the airplane center of gravity is assumed.

The equations for nose-gear yaw loads are shown as specified in FAR 25.499(a):

$$V_{NG} = n_z(V_{NG})_s \qquad (7.59)$$

$$D_{NG} = 0 \qquad (7.60)$$

$$S_{NG} = \pm0.80V_{NG} \qquad (7.61)$$

where $(V_{NG})_s$ is the static condition defined by Eq. (7.5), and $n_z = 1.0$ limit.

7.6.4 Unsymmetrical Braking per FAR/JAR 25.499(b)

With the airplane assumed to be in static equilibrium with the loads resulting from application of main gear brakes on one side of the airplane, the nose gear, its attaching structure, and the fuselage structure forward of the airplane center of gravity must be designed for the following loads.

Nose gear loads:

$$V_{NG} = n_z W[B + E\mu_{MG}(0.50 + BL_{cg}/T)]/(C + 0.50E\mu_{MG}) \qquad (7.62)$$

$$D_{NG} = 0 \qquad (7.63)$$

$$S_{NG} = D_{MGr}(BL_{cg} - 0.50T)/C < 0.80V_{NG} \tag{7.64}$$

Main gear loads:

$$V_{MGr} = n_z W(0.50 + BL_{cg}/T) - 0.50V_{NG} \tag{7.65}$$

$$V_{MGl} = n_z W - V_{MGr} - V_{NG} \tag{7.66}$$

$$D_{MGr} = \mu_{MG} V_{MGr} \tag{7.67}$$

$$D_{MGl} = 0 \tag{7.68}$$

$$S_{MGr} = -0.50S_{NG} \tag{7.69}$$

$$S_{MGl} = 0.50S_{NG} \tag{7.70}$$

where the equations shown are for braking on the right gears.

The load factors applied to the airplane for this condition are specified in the regulations:

$$n_z = 1.0 \text{ limit}$$

$$n_y = 0$$

The forward acting load factor may be determined from the drag load as calculated from Eq. (7.67):

$$n_x = \mu_{MG} V_{MGr}/W \tag{7.71}$$

where V_{MGr} is the vertical load on the main gear that has the applied braking force (lb), W is the airplane maximum taxi weight (lb), and $\mu_{MG} = 0.80$ for normal tire conditions, except where main gear brakes are torque limited, a reduced forward acting load factor at the airplane center of gravity may be used.

Side and vertical loads at the ground contact point on the nose gear are as required for equilibrium. The ratio of the nose gear side load to vertical load does not need to exceed 0.80.

7.6.5 Nose Gear Steering per FAR/JAR 25.499(e)

With the airplane at maximum taxi weight and the nose gear in any steerable position, the combined application of full normal steering torque and a vertical force equal to the maximum static reaction on the nose gear must be considered in designing the nose gear, its attaching structure, and the forward fuselage structure.

The static nose gear limit vertical load is defined by Eq. (7.5).

7.6.6 Pivoting per FAR/JAR 25.503

The airplane is assumed to pivot about the main gear on one side with the brakes on that side locked. The airplane is assumed to be in static equilibrium, with the loads being applied at the ground contact points as shown in Fig. 7.2.

Nose gear loads:

$$V_{NG} = n_z(V_{NG})_s \tag{7.72}$$

$$D_{NG} = S_{NG} = 0 \tag{7.73}$$

Main gear loads:

$$V_{MGr} = n_z(V_{MGr})_s \tag{7.74}$$

$$V_{MGl} = n_z(V_{MGl})_s \tag{7.75}$$

$$D_{MGr} = D_{MGl} = 0 \tag{7.76}$$

$$S_{MGr} = S_{MGl} = 0 \tag{7.77}$$

where s is the static condition defined by Eqs. (7.1), (7.2), and (7.5), and $n_z = 1.0$ limit.

The torque about the locked main gear is defined as

$$\text{torque pivot} = V_{MGr} \mu_{MG} K_{piv} L_{piv} \tag{7.78}$$

where $\mu_{MG} = 0.80$, and for a two-wheel gear

$$L_{piv} = 0.50F \tag{7.79}$$

and for a four-wheel gear

$$L_{piv} = 0.50(F^2 + d^2)0.50 \tag{7.80}$$

and where $K_{piv} = 1.33$, F is the distance between wheels on the same axle (in.), and d is the distance between axles on the same gear (in.).

7.7 Towing Conditions

Structural strength requirements for towing the airplane are discussed in this section. Design tow loads are specified in the regulations as a function of the airplane design gross weight and the direction of the tow. These loads are not affected by airplane center of gravity position, which will contribute only to the magnitude of the vertical load applied to the landing gears for a specific tow condition.

7.7.1 Towing Loads per FAR/JAR 25.509

The towing loads F_{TOW} are applied parallel to the ground at the landing gear towing fittings as shown in Fig. 7.13. The requirements as summarized in Table 7.6 are specified in FAR/JAR 25.509 as a function of airplane maximum design taxi (ramp) gross weight.

The tow loads applied to each main gear unit are $0.75 F_{TOW}$ in the directions noted in FAR 25.509.

The tow loads applied to the nose gear (or tail wheel) are $1.0 F_{TOW}$ or $0.50 F_{TOW}$ depending on the direction of the tow. The tow loads applied to the landing gear tow fittings must be reacted as noted in FAR/JAR 25.509(c).

For tow points not on the landing gears the requirements of FAR/JAR 25.509(b) should be considered.

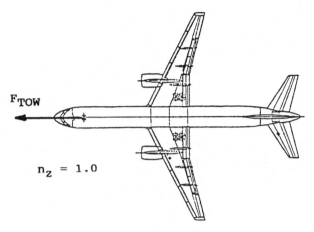

$$F_{\text{TOW}}$$

$$n_z = 1.0$$

Fig. 7.13 Towing condition.

7.8 Jacking Loads per FAR/JAR 25.519

In February 1993 jacking load requirements were proposed to be incorporated into FAR 25.519 as part of the FAR/JAR harmonization process. These requirements set forth the design factors applied to the static ground load conditions for the most critical combination of gross weight and airplane center of gravity position. The maximum allowable limit load at each jack pad must be specified.

The aircraft structure must be designed for jacking by the landing gears at the design maximum taxi weight (MTW) as specified in Table 7.7.

When jacking by other airplane structures is allowed, the maximum jacking weight (MJW) must be specified. The design load requirements are also specified in Table 7.7.

7.9 Tethering Problem

The tethering problem has been a concern of many airlines that operate aircraft in the Pacific typhoon regions. It has also concerned European operators who fly into mountainous airports where high winds are common.

Several names have been used over the years to describe this condition: 1) mooring, 2) tethering, 3) picketing (early JAR proposal), and 4) tie down (current harmonization proposal).

Table 7.6 Gear load equations for towing
conditions per FAR 25.509 with F_{TOW}
applied at tow fittings

W,[a] lb	n_z[b]	F_{TOW},[c] lb limit
$< 30,000$	1.0	$0.30W$
30,000–100,000	1.0	$(6W + 450,000)/70$
$> 100,000$	1.0	$0.15W$

[a] W is the maximum taxi (ramp) weight (lb).
[b] $n_z = 1.0$ (limit) acting at the airplane center of gravity.
[c] F_{TOW} is applied parallel to the ground as defined in FAR 25.509.

Table 7.7 Jacking loads per FAR/JAR 25.219

	Gross weight	Vertical load	Horizontal load applied in any direction
Loads applied at landing gears			
	MTW	$1.33SFV_s$	0
	MTW	$1.33SFV_s$	$0.33SFV_s$
Loads applied to other jack points			
Airplane	MJW	$1.33SFV_{static}$	0
structure	MJW	$1.33SFV_{static}$	$0.33SFV_{static}$
Jack pads	MJW	$2.0SFV_{static}$	0
and local	MJW	$2.0SFV_{static}$	$0.33SFV_{static}$
structure			

Note: $SF = 1.5$ for ultimate loads; V_s is the static loads as calculated from Eqs. (7.1), (7.2), and (7.5); and V_{static} is the vertical static reaction at each jacking pad for the selected fore and aft center of gravity positions at the maximum approved jacking weight (MJW).

The FAR/JAR harmonization proposal requires that if tie-down points are provided, the main tie-down points and local structure must withstand the limit loads resulting from a 65-kn horizontal wind, applied from any direction.

Some airline requirements stipulate the tie-down points be investigated for winds up to 100 kn. If an aircraft cannot be flown out of a typhoon area, the airplane must be tied down in such a manner that the nose of the aircraft faces into the wind. As the wind changes direction, the aircraft must be moved accordingly. Reports from Pacific operators have indicated that some of the very large jets have "jumped around" while weathervaning in a tied-down condition.

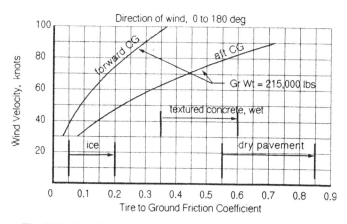

Fig. 7.14 Steady wind velocity required for weathervaning.

7.9.1 Steady Wind Velocity Required for Weathervaning

An example of the steady wind velocity required for weathervaning is shown in Fig. 7.14. These curves indicate the wind velocities that may become a concern of an operator, depending on the airplane gross weight, center of gravity position, and the type of surface on which the aircraft is parked.

Reference

[1]Anon., "Ground Loads," ANC-2 Bulletin, U.S. Depts. of the Air Force, Navy, and Commerce, Civil Aeronautics Adminstration, Oct. 1952.

8
Horizontal Tail Loads

Horizontal tail loads affect the design of a significant part of the aircraft structure and hence require careful consideration of the various design requirements and resulting conditions. In general the structures that are designed by horizontal tail loads are 1) the horizontal tail stabilizer and elevator, 2) the body structure aft of the pressure bulkhead and horizontal tail support structure, 3) the aft fuselage monocoque structure, 4) the fuselage center section (overwing) structure, and 5) the stabilizer actuator (jackscrew mechanism).

The geometric parameters and sign convention for horizontal tail loads are shown in Fig. 8.1.

8.1 Horizontal Tail Design Load Envelopes

Before discussing the methods of analysis for calculating horizontal tail loads, the critical load conditions forming the design load envelope will be considered. A typical horizontal tail design load envelope is shown in Fig. 8.2 with total tail airload L_t plotted as a function of total tail pitching moment M_t due to airload. The purpose of this graph is to provide a means of selecting the critical design loads for the horizontal tail and aft body structure. The critical loads for a given type of symmetrical flight condition may be determined from the maximum or minimum loads for each type selected.

A similar set of graphs of shear or bending moment vs torsion may be plotted at several selected spanwise stations, from which the critical conditions for a given type of condition may be determined.

8.2 Balanced Maneuver Analysis

Balanced maneuver conditions were discussed in Chapter 2 with respect to the requirements and development of the steady-state symmetric flight equations. In this section a more detailed discussion of the application of these requirements in calculating horizontal tail loads and some historical background will be presented.

8.2.1 Historical Perspective

During the certification of the 707 airplane in 1956 and the DC-8 in 1957, consideration was given to mistrim stabilizer conditions due to the movable horizontal stabilizer. Before 1952 commercial transports had fixed stabilizers with trim about the pitch axis being obtained through the elevator system. The FAA regulations were changed by Amendment 25-23 in April 1970 to include consideration of in-trim and out-of-trim flight conditions [see FAR 25.331(a)(4)].[1]

The method of analysis for determining the in-trim and out-of-trim conditions is presented in Sec. 8.2.6 and has not changed significantly over the years.

a)

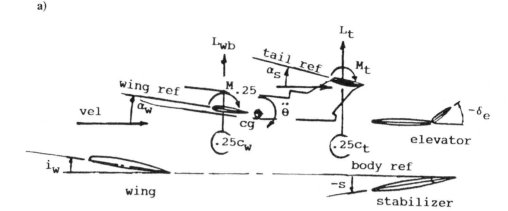

b)

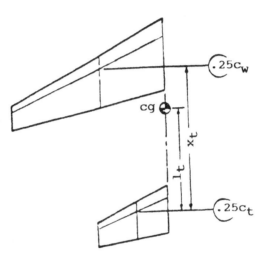

Fig. 8.1 Geometric parameters and sign convention for horizontal tail loads: a) sign convention and b) geometric parameters.

8.2.2 Balancing Tail Load Equations

The relationship between horizontal tail load, pitching moment, stabilizer angle of attack, and elevator angle will be considered for a steady-state maneuver in which pitching acceleration is zero.

The equations for horizontal tail load and pitching moment are shown in Eqs. (8.1) and (8.2) as a function of the stabilizer angle of attack and elevator angle:

$$L_t = L_{\alpha s}\alpha_s + L_{\delta e}\delta_e + L_c \tag{8.1}$$

$$M_t = M_{\alpha s}\alpha_s + M_{\delta e}\delta_e + M_c \tag{8.2}$$

where L_t is the horizontal tail load (lb), M_t is the horizontal tail pitching moment about the 0.25 mac$_t$ (in.-lb), $L_{\alpha s}$ and $M_{\alpha s}$ are the tail load and pitching moment due to unit α_s (lb/deg and in.-lb/deg, respectively), $L_{\delta e}$ and $M_{\delta e}$ are the tail load

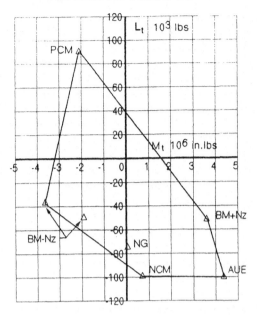

Fig. 8.2 Typical design load envelope horizontal tail centerline airloads: AUE—abrupt-up elevator; BM+Nz—balanced maneuver at $+n_z$; BM+Nz—balanced maneuver at $-n_z$; NG—static negative gust; NCM—negative checked maneuver; PCM—positive checked maneuver.

and pitching moment due to unit δ_e (lb/deg and in.-lb/deg, respectively), L_c and M_c are the tail load and pitching moment due to unit built-in camber (lb and in.-lb, respectively), α_s is the horizontal stabilizer angle of attack (deg), and δ_e is the elevator angle (deg).

The horizontal stabilizer angle of attack α_s as used in Eqs. (8.1) and (8.2) may be defined as a function of wing angle of attack, the curvilinear flight increment due to pitching velocity, and fuselage flexibility terms plus the initial trim setting of the stabilizer:

$$\alpha_s = (1 - \epsilon_{\alpha w})\alpha_w - \epsilon_0 + 57.3 l_t g (n_z - 1)/V_t^2$$
$$+ s + \left(\frac{d\alpha_s}{dn_z}\right) n_z + \left(\frac{d\alpha_s}{dL_t}\right) L_t + \left(\frac{d\alpha_s}{dM_t}\right) M_t \tag{8.3}$$

The increment due to wing angle of attack is modified by the wing downwash at the horizontal stabilizer:

$$\blacktriangle\alpha_s = \alpha_w - \epsilon_{\alpha w}\alpha_w - \epsilon_0$$
$$= (1 - \epsilon_{\alpha w})\alpha_w - \epsilon_0 \qquad \text{(deg)} \tag{8.4}$$

where $\epsilon_{\alpha w}$ is the change in wing angle of attack due to downwash at the horizontal tail reference point (usually the quarter-chord of the horizontal tail mean aerodynamic chord), and ϵ_0 is the downwash angle at the horizontal tail at $\alpha_w = 0$ (deg).

The curvilinear flight increment due to pitching velocity may be derived for a steady-state maneuver from Eq. (8.5):

$$\Delta\alpha_s = 57.3\dot{\theta}l_t/V_t \qquad \text{(deg)} \qquad (8.5)$$

$$l_t = x_t + (0.25 - cg)c_w \qquad \text{(ft)} \qquad (8.6)$$

The pitching velocity $\dot{\theta}$ shown in Eq. (8.5) is defined in Eq. (8.7) for a steady-state maneuver:

$$\dot{\theta} = g(n_z - 1)/V_t \qquad \text{(rad/s)} \qquad (8.7)$$

The stabilizer setting s is with respect to a selected wing reference plane, or if a body reference is used,

$$s = s_{\text{body}} - i_{\text{wing}} \qquad (8.8)$$

where i_{wing} is the wing reference plane angle of incidence (deg).

The final three terms within the parentheses represent the change in stabilizer angle of attack due to body flexibility as defined in Fig. 8.3.

Using the simplification derived in Sec. 2.3.1, the relationship between horizontal tail load, pitching moment, and the airplane balance requirements for a steady-state maneuver may be derived:

$$BTL = L_t - M_t/x_t \qquad (2.15)$$

a)

b)

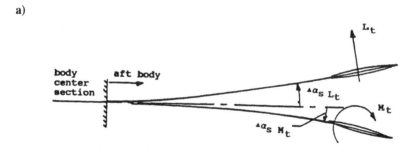

Fig. 8.3   Vertical bending due to body flexibility: a) bending due to horizontal tail loads and b) bending due to load factor and pitching acceleration.

where BTL is the balancing tail load calculated using the procedure discussed in Sec. 2.3.1, assuming $M_t = 0$ lb.

8.2.2.1 Equations for conditions with unknown elevator angle.

Equations (8.1–8.3) and (2.15) may now be rearranged and represented in matrix form as shown in Eq. (8.9) for conditions where the elevator angle is an unknown. Simultaneous solution of this set of equations will give the tail load, pitching moment, stabilizer angle of attack, and the elevator required to accomplish the maneuver. For solution of these equations the stabilizer geometric trim position is known,

$$
\begin{bmatrix}
1 & 0 & -L_{\alpha s} & -L_{\delta e} \\
0 & 1 & -M_{\alpha s} & -M_{\delta e} \\
x_t & -1 & 0 & 0 \\
b_1 & b_2 & 1 & 0
\end{bmatrix}
\begin{bmatrix}
L_t \\
M_t \\
\alpha_s \\
\delta_e
\end{bmatrix}
=
\begin{bmatrix}
L_c \\
M_c \\
BTLx_t \\
c_1
\end{bmatrix}
\tag{8.9}
$$

where

$$
b_1 = -\left(\frac{d\alpha_s}{dL_t}\right) \tag{8.10}
$$

$$
b_2 = -\left(\frac{d\alpha_s}{dM_t}\right) \tag{8.11}
$$

$$
c_1 = (1 - \epsilon_{\alpha w})\alpha_w - \epsilon_0 + 57.3 l_t g(n_z - 1)/V_t^2 + s_{\text{trim}} + \left(\frac{d\alpha_s}{dn_z}\right)n_z \tag{8.12}
$$

where s_{trim} is the stabilizer geometric angle with respect to fuselage reference axis (deg) [see Eq. (8.8)].

8.2.2.2 Equations for conditions with known elevator angle.

For steady-state maneuvers where the elevator angle is known such as the case where the elevator is blowdown limited, Eq. (8.9) can be reduced to three unknowns as shown in Eq. (8.13). Simultaneous solution of this set of equations will give the tail load, pitching moment, and stabilizer angle of attack. The stabilizer geometric position required to attain the load factor n_z may be calculated from Eq. (8.14):

$$
\begin{bmatrix}
1 & 0 & -L_{\alpha s} \\
0 & 1 & -M_{\alpha s} \\
x_t & -1 & 0
\end{bmatrix}
\begin{bmatrix}
L_t \\
M_t \\
\alpha_s
\end{bmatrix}
=
\begin{bmatrix}
L_c + L_{\delta e}\delta_e \\
M_c + M_{\delta e}\delta_e \\
BTLx_t
\end{bmatrix}
\tag{8.13}
$$

where

$$
s = \alpha_s - \left[(1 - \epsilon_{\alpha w})\alpha_w - \epsilon_0 + 57.3 l_t g(n_z - 1)/V_t^2 \right.
$$
$$
\left. + \left(\frac{d\alpha_s}{dn_z}\right)n_z + \left(\frac{d\alpha_s}{dL_t}\right)L_t + \left(\frac{d\alpha_s}{dM_t}\right)M_t \right] \tag{8.14}
$$

Consideration will now be given to the determination of the stabilizer trim settings for a level flight condition and for the special case where the elevator angle is limited by the maximum capability of the control system such as blowdown at high airspeeds.

8.2.3   Determination of Initial Flight Condition at $n_z = 1.0$

Two scenarios for calculating the initial flight trim settings will be considered depending on whether the aircraft has a movable stabilizer or whether trim is accomplished with the elevators.

8.2.3.1   Stabilizer trim required with elevator at neutral.
The elevator is assumed to be zero or at the neutral point setting depending on the downrig or uprig built into the control system. Equation (8.13) is solved for tail load, pitching moment, and stabilizer angle of attack with the airplane assumed in level flight at $n_z = 1.0$. The stabilizer trim setting s_1 is then calculated from Eq. (8.14).

8.2.3.2   Elevator required for trim.
If the stabilizer position is known, tail load, pitching moment, stabilizer angle of attack, and the elevator required for trim may be calculated from the simultaneous solution of Eq. (8.9) where $n_z = 1$.

8.2.4   Solution of Equations for Elevator Required at Load Factor

Simultaneous solution of Eq. (8.9) will give the horizontal tail load, pitching moment, stabilizer angle of attack, and elevator angle for a steady-state maneuver at load factor n_z. For aircraft with movable stabilizers, three scenarios may be considered depending on how the maneuver is flown: by elevator, stabilizer, or a combination of both.

8.2.4.1   Maneuver with the airplane trimmed for level flight.
With the airplane trimmed at $n_z = 1.0$ as discussed in Sec. 8.2.3, the maneuver at load factor n_z may be accomplished using elevator, stabilizer, or a combination of both depending on the design configuration of the airplane control system.

For conditions where the maneuver is flown using the movable stabilizer only, simultaneous solution of Eq. (8.13) will give the tail load, pitching moment, and stabilizer angle of attack. The stabilizer position required to attain the desired load factor n_z is then calculated from Eq. (8.14). Usually this condition is not critical due to the low horizontal tail torsion when compared with the condition in which the maneuver is accomplished with the elevator.

For conditions in which the maneuver is flown using the elevator only, solution of Eq. (8.9) will give the tail load, pitching moment, stabilizer angle of attack, and the elevator angle required to attained the load factor n_z.

8.2.5   Balancing Tail Load Analysis for Elevator-Limited Conditions

The special condition where the maximum available elevator is limited by blowdown or control system stops must be considered. Solution of Eq. (8.13) will result in horizontal tail load, pitching moment, and stabilizer angle of attack for conditions at δ_{emax}. If the design maneuver load factor n_z is not attainable at the elevator-limited condition, the stabilizer geometric angle required to accomplish the maneuver may be calculated from Eq. (8.14). An example of this type of condition is shown in Fig. 8.4.

Elevator-limited conditions are shown in Fig. 8.4 at two dive speeds. The design maneuver load factor, $n_z = 2.5$, is obtainable for Mach $= 0.90$ with the airplane trimmed for $n_z = 1.0$ flight with zero elevator, whereas the Mach $= 0.95$ conditions require use of a stabilizer to obtain the design maneuver load factor.

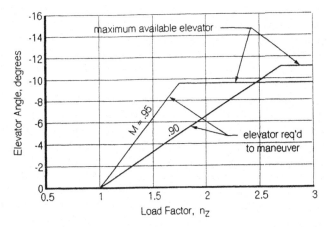

Fig. 8.4 Elevator required to maneuver at design dive speeds.

8.2.6 Out-of-Trim Considerations for Balanced Maneuvers

As stated in FAR/JAR 25.331(a)(4), out-of-trim flight conditions specified in FAR/JAR 25.255 must be considered as follows: 1) FAR/JAR 25.255(a)(1)—a 3-s movement of the longitudinal trim system at its normal rate, and 2) FAR/JAR 25.255(a)(2)—the maximum mistrim that can be sustained by the autopilot while maintaining level flight in the high-speed cruising condition.

Maximum trim rates are shown in Table 8.1 for two aircraft with movable stabilizers. The 3-s runaway stabilizer increments are also shown.

8.2.6.1 Historical perspective.
Several of the jet transports certified before 1968 had elevator-limited (blowdown) conditions at design airspeeds M_D. For these aircraft the stabilizer was assumed to be retrimmed as necessary to produce the design load factor of $n_z = 2.5$ (or zero). This maximized the horizontal tail torsion and the stabilizer actuator design requirements.

Many aircraft designed after 1968 were able to attain the design maneuver load factor, $n_z = 2.5$, at elevator angles less than limited by blowdown. For these aircraft consideration was given to the 3-s out-of-trim requirements of FAR 25.255. An example of out-of-trim condition is shown in Fig. 8.5 for a steady-state balanced maneuver using the 3-s rule.

The elevator angles required to maneuver at design dive speeds are shown in this figure at two Mach numbers. Each condition is trimmed with the stabilizer such

Table 8.1 Stabilizer normal mistrim rates for a typical jet transport

Airplane	Airspeeds and flap position	Stabilizer rate, deg/s	Δs, 3-s deg
A	Flaps down	±0.60	±1.8
	Flaps up	±0.20	±0.6
B	Flaps down	±0.50	±1.5
	at V_C	±0.36	±1.1
	at V_D	±0.33	±1.0

a)

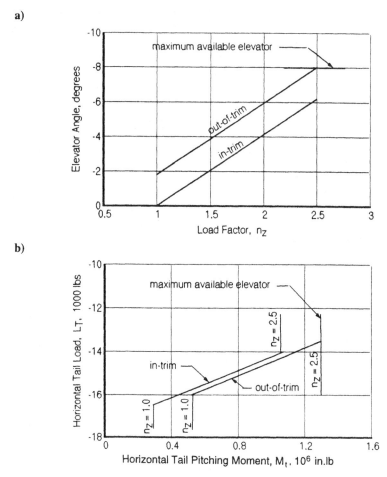

b)

Fig. 8.5   Out-of-trim conditions for steady state maneuvers: a) elevator angle vs load factor and b) horizontal tail load vs pitching moment.

that the elevator angle is zero for level flight. As noted the elevator angle at Mach = 0.95 is blowdown limited such that the design load factor of $n_z = 2.5$ cannot be attained without retrimming the stabilizer. For this aircraft the assumption was made that the remainder of the pull-up to design load factor was made with the stabilizer.

8.3   Abrupt Unchecked Elevator Conditions

Abrupt unchecked elevator maneuvers were discussed in Sec. 2.5 with respect to the requirements and development of the equations of motion. In this section a more detailed discussion of the application of these requirements in calculating horizontal tail loads and some historical background will be presented.

Horizontal tail loads may be calculated from modifications of Eqs. (8.1) and (8.2) as shown in Eqs. (8.15) and (8.16):

$$L_t = L_{t n z=1} + L_{\alpha s} \Delta\alpha_s + L_{\delta e} \Delta\delta_e \qquad (8.15)$$

$$M_t = M_{t n z=1} + L_{\alpha s} \Delta\alpha_s + M_{\delta e} \Delta\delta_e \qquad (8.16)$$

where $L_{tnz=1}$ and $M_{tnz=1}$ are the horizontal tail load and pitching moment for an initial 1-g flight condition (lb and in.-lb, respectively), $L_{\alpha s}$ and $M_{\alpha s}$ are the horizontal tail load and pitching moment due to change in α_s (lb/deg and in.-lb/deg, respectively), $L_{\delta e}$ and $M_{\delta e}$ are the horizontal tail load and pitching moment due to change in δ_e (lb/deg and in.-lb/deg, respectively), $\Delta\alpha_s$ is the change in horizontal tail angle of attack due to airplane response and body flexibility (deg), and $\Delta\delta_e$ is the change in elevator angle (deg).

Consideration will be given to simplified procedures where the horizontal tail loads due to an abrupt unchecked elevator can be calculated, without using a complete time history analysis as discussed in Sec. 2.4.

8.3.1 Historical Perspective

Before 1952, horizontal tail loads were obtained for the maximum up elevator unchecked maneuver as determined from two-thirds maximum pilot effort at the condition being investigated, provided that elevator hinge moments were based on reliable data. This is the same as stated in the current regulations, FAR 25.331(c)(1), which refer to FAR 25.333(b). The amount of pilot effort required for design purposes would then be 200 lb for an airplane with a control wheel.

The maximum up elevator available, limited by pilot effort, may vary with airplane center of gravity position. For example, on the Boeing 377 Stratocruiser, at the forward center of gravity, 10-deg up elevator was available, whereas at the aft center of gravity, 15-deg up elevator was used for structural analysis.

With the advent of commercial jet aircraft using sophisticated control systems, i.e., elevator tabs and balance panels or, as is the case for modern jets, fully powered surfaces using hydraulic power control units, maximum available elevator conditions were not limited by control force but rather by the maximum hinge moment available from the control system. Determination of the maximum elevator available, as discussed in Sec. 2.5.2, is shown graphically in Fig. 12.2.

During initial certification of some commercial transports before 1960, horizontal tail loads were computed assuming an instantaneous application of elevator, neglecting airplane response, such that $\Delta\alpha_s = 0$ in Eqs. (8.16) and (8.17). Thus the resulting horizontal tail loads and pitching moments were calculated on a conservative basis, being a function of only the initial level flight tail load and the increment due to elevator displacement. These abrupt maneuvers were called "instantaneous elevator conditions."

For certification of aircraft after 1960, including growth versions of the earlier airplanes, airplane response was considered.

8.3.2 Horizontal Tail Loads Due to Abrupt Elevator Input

With limited capability of solving the equations of motion discussed in Sec. 2.4.1, an alternate procedure was used to determine horizontal tail loads due to abrupt elevator inputs.

The incremental tail load and pitching moment due to abrupt application of the elevator may be written as follows:

$$\Delta L_{t\theta} = L_{\alpha s}\Delta\alpha_s + L_{\delta e}\Delta\delta_{e\,\text{max}} \tag{8.17}$$

$$\Delta M_{t\theta} = M_{\alpha s}\Delta\alpha_s + M_{\delta e}\Delta\delta_{e\,\text{max}} \tag{8.18}$$

where $\Delta L_{t\theta}$ and $\Delta M_{t\theta}$ are the horizontal tail load and pitching moment due to change in elevator and airplane response (lb and in.-lb, respectively).

The airplane response factor for an abrupt unchecked maneuver may be defined as shown in Eq. (8.19):

$$k_r = \blacktriangle L_{t\theta}/(L\blacktriangle\delta_{e\max}) \tag{8.19}$$

After substitution of Eq. (8.19) into Eq. (8.17), the change in stabilizer angle of attack becomes

$$\blacktriangle\alpha_s = (k_r - 1)(L_{\delta e}/L_{\alpha s})\blacktriangle\delta_{e\max} \tag{8.20}$$

Inserting Eq. (8.15) into Eq. (8.20), one can define the net horizontal tail load due to an abrupt unchecked elevator in Eq. (8.21) in terms of the response parameter k_r:

$$L_t = L_{tnz=1} + k_r L_{\delta e}\blacktriangle\delta_{e\max} \tag{8.21}$$

Horizontal tail pitching moment may be calculated from Eq. (8.16) using the stabilizer angle of attack computed from Eq. (8.20) and $\blacktriangle\delta_{e\max}$.

8.3.3 Horizontal Tail Analysis Load Surveys

For analysis load surveys to determine the critical abrupt elevator horizontal tail load condition, a conservative response factor of $k_r = 0.90$ may be used.

Typical response parameters k_r shown in Fig. 8.6 for a commercial jet transport are calculated from the time history analyses discussed in Sec. 2.5 using ramped elevator inputs. Examples for an abrupt-up unchecked elevator are shown in Tables 2.6a and 2.6b.

The response factor k_r will be further reduced if the exponential elevator input discussed in Sec. 2.5.1 is used to solve the equations of motion. Examples of these analyses for an abrupt-up unchecked elevator using the exponential elevator motion are shown in Tables 2.7a and 2.7b.

The question may be raised, why use an assumed response factor when the time history analysis may be readily solved on a personal computer? The use of the response factor is recommended only for load surveys where the critical speed/altitude condition may be determined.

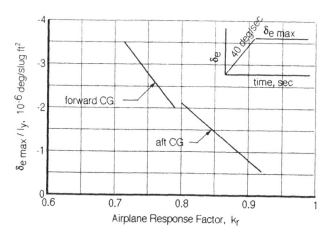

Fig. 8.6 Response factor k_r for abrupt elevator maneuvers.

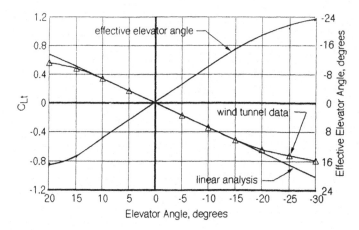

Fig. 8.7   Elevator nonlinear characteristics.

8.3.4   Nonlinear Elevator Characteristics

Equations (8.16) and (8.17) are based on linearized aerodynamic parameters. The use of nonlinear elevator characteristics needs to be considered in computing tail loads at maximum elevator to reduce unneeded structural conservatism.

From wind-tunnel data the variation of tail load due to elevator angle may be calculated as shown in Fig. 8.7, where an effective elevator angle may be determined from the ratio of the actual tunnel data to the linearized analysis representation of the variation of horizontal tail load with elevator angle:

$$\delta_{eff} = \blacktriangle C_{Lt}\delta_{actual}/\blacktriangle C_{Lt\,linear} \tag{8.22}$$

where $\blacktriangle C_{Lt}$ is the incremental lift coefficient due to elevator obtained from wind-tunnel data, $\blacktriangle C_{Lt\,linear}$ is the linearized incremental lift coefficient due to elevator based on a rigid tail analysis, and δ_{actual} is the elevator angle as measured in the wind tunnel (deg).

A similar curve may be determined for horizontal tail pitching moment. A conservative effective elevator angle may then be used for structural analysis based on both the lift or pitching moment due to elevator. In general, elevator characteristics are usually assumed linear from ±15-deg elevator depending on the camber built into the horizontal tail and the Mach number under consideration. A conservative analysis may be used when the horizontal tail lift due to elevator varies with stabilizer angle of attack.

The effect of nonlinear elevator characteristics is shown in Table 8.2. The initial tail loads and elevator trim angle at $n_z = 1.0$ have been assumed equal to zero for convenience of analysis. The inclusion of nonlinear effects results in a 9% reduction in horizontal tail load and a 5% reduction in horizontal tail pitching moment for this example.

Consideration of nonlinear effects for solving the equations of motion in Sec. 2.4.1 may be accomplished by describing the effective elevator angle variation with time instead of the actual elevator angle, as noted in Fig. 8.8.

8.4   Checked Maneuver Conditions

Abrupt checked maneuvers were discussed in Sec. 2.6 with respect to the requirements and development of the equations of motion. In this section a more

Table 8.2   Horizontal tail loads due to abrupt-up elevator,
effect of nonlinear characteristics[a]

Altitude, ft	13,400	2,000	0
V_e, keas	205	255	280
Mach	0.31	0.386	0.423
$L_{\alpha s}$, lb/deg	4,082	6,193	7,434
$L_{\delta s}$, lb/deg	2,403	3,532	4,127
$M_{\alpha s}$, in.-lb/deg	−63,400	−93,600	−109,500
$M_{\delta s}$, in.-lb/deg	−108,900	−164,800	−194,000
$\delta_{e\max}$, deg	−30.0	−20.0	−16.0
$\delta_{e\text{eff}}$, deg	−23.0	−18.5	−16.0
$\alpha_{s\text{lin}}$, deg	1.766	1.141	0.888
$\alpha_{s\text{non}}$, deg	1.354	1.055	0.888
Linear analysis			
$\quad L_t$, lb	−64,881	−63,576	−59,429
$\quad M_t$, 10^6 in.-lb	3.155	3.189	3.007
Nonlinear analysis			
$\quad L_t$, lb	−49,742	−58,808	−59,429
$\quad M_t$, 10^6 in.-lb	2.419	2.950	3.007

[a]Horizontal tail loads are shown using both linear and nonlinear elevator characteristics. Loads are calculated using Eq. (8.21) assuming the response factor $k_r = 0.90$. For this example the initial parameters at $n_z = 1.0$ are assumed equal to zero. All loads shown are limit.

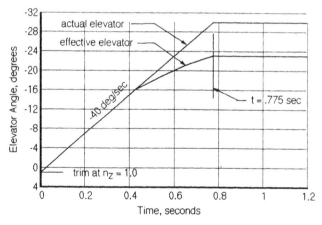

Fig. 8.8   Unchecked elevator time history for nonlinear analysis.

detailed discussion of the application of these requirements in calculating horizontal tail loads and some historical background will be presented.

8.4.1 Historical Perspective

In the days before commercial jet transports, horizontal tail loads were computed based on a prescribed pitching acceleration based on the minimum requirements in the early versions of FAR 25[1] and its predecessor, CAR 4b.[2] These requirements are still shown in the regulations as noted in Eqs. (7.1) and (7.2). In general, the minimum requirements produce horizontal tail loads that are less than those obtained from a rational analysis except for aircraft with very large pitching moments of inertia, such as the 747 airplanes.

The horizontal tail load and pitching moment may be calculated in Eqs. (8.23) and (8.24) for the prescribed pitching acceleration:

$$L_t = L_{tnz} + L_{t\ddot{\theta}} \tag{8.23}$$

$$M_t = L_{tnz} + M_{t\ddot{\theta}} \tag{8.24}$$

where L_{tnz} is the balancing tail load at a load factor of n_z, M_{tnz} is the horizontal tail pitching moment at load factor of n_z consistent with the balancing tail load, $L_{t\ddot{\theta}}$ is the incremental tail load due to pitching acceleration, and $M_{t\ddot{\theta}}$ is the incremental tail pitching moment load due to pitching acceleration.

The positive pitch acceleration requirements at $n_z = 1.0$ are defined by Eq. (2.37), and the negative pitch acceleration at design load factor n_z is defined by Eq. (2.38). The incremental horizontal tail load and pitching moment are then related to pitching acceleration as shown in Eq. (8.25):

$$I_y\ddot{\theta} = M_{t\ddot{\theta}} - l_t L_{t\ddot{\theta}} \tag{8.25}$$

where I_y is the airplane moment of inertia about the airplane center of gravity (slug ft²), and l_t is the distance from horizontal tail load reference axis to the airplane center of gravity (ft).

Neglecting the horizontal tail pitching moment term as small, one can calculate the net horizontal tail load for a checked maneuver from Eq. (8.26):

$$L_t = L_{tnz} - I_y\ddot{\theta}/l_t \tag{8.26}$$

By using the definition of the incremental tail load and pitching moment due to pitching acceleration shown in Eqs. (8.17) and (8.18), one can show that it is conservative to neglect the pitching moment in the calculation of tail load using Eq. (8.26).

8.4.2 Checked Maneuver Tail Loads Using a Rational Approach

It became evident from studies made on the 707 airplane that a more rational approach must be made in determining horizontal tail loads in checked maneuvers. Not having the sophistication of modern digital computers, engineers computed checked maneuver tail loads using the following approach.

The assumption was made that the elevator was instantaneously checked back to neutral or overchecked a specified amount depending on the airplane center of gravity position, while the airplane was in a steady-state maneuver at the design load factor. These criteria were developed based on the military requirements discussed in Sec. 2.7.

8.4.2.1 Positive checked maneuvers, forward center of gravity positions, and all negative checked maneuvers. For positive maneuvers with forward center of gravity positions and all negative maneuvers, the elevator was returned to neutral from the elevator required to produce the steady-state balancing tail load at design load factor n_z. The amount of checked elevator was defined as shown in Eq. (8.27):

$$\blacktriangle\delta_{ecm} = -\blacktriangle\delta_{ebal} \tag{8.27}$$

where $\blacktriangle\delta_{ecm}$ is the incremental checked maneuver elevator from the elevator position required for the steady-state balanced maneuver at the load factor under consideration.

8.4.2.2 Positive checked maneuvers for aft center of gravity positions. For positive maneuvers with aft center of gravity positions, the elevator was returned in the opposite direction to 50% of the elevator required to produce the steady-state balancing tail load at design load factor n_z. The amount of checked elevator was defined as shown in Eq. (8.28):

$$\blacktriangle\delta_{ecm} = -1.5\blacktriangle\delta_{ebal} \tag{8.28}$$

The basic problem with the criteria for checked maneuvers at the aft position is defining what the aft center of gravity is. As growth is considered in a given commercial transport, the aft center of gravity is usually cut off as shown in Fig. 14.2. This causes the amount of checked elevator to increase as the center of gravity moves forward.

The rationale for this condition in the military criteria is that the airplanes (namely fighters) have very low stability at the aft center of gravity positions. The criteria specifically apply to an unstable airplane. Commercial transport aircraft by definition are stable within the total center of gravity positions allowed for operation in the airplane flight manuals; hence the criteria discussed in this section are very conservative.

8.4.2.3 Checked maneuver analysis using simplified approach. Neglecting any airplane response during the checkback of the elevator from the steady-state maneuver, one can determine the pitching acceleration from Eq. (8.29):

$$\ddot{\theta} = (M_{t\delta e} - l_t L_{t\delta e})\blacktriangle\delta_{ecm}/I_y \qquad \text{(rad/s)} \tag{8.29}$$

The total horizontal tail load and pitching moment due to the checked maneuver can be written in terms of the balance increment plus the increment due to pitching acceleration caused by returning the elevator to neutral or the overchecked position:

$$L_{tcm} = L_{tbal} + L_{t\delta e}\blacktriangle\delta_{ecm} \qquad \text{(lb)} \tag{8.30}$$

$$M_{tcm} = M_{tbal} + M_{t\delta e}\blacktriangle\delta_{ecm} \qquad \text{(in.-lb)} \tag{8.31}$$

where L_{tcm} and M_{tcm} are the check maneuver tail load and pitching moment (lb and in.-lb, respectively), L_{tbal} and M_{tbal} are the balancing tail load and pitching moment at load factor n_z (lb and in.-lb, respectively), $L_{t\delta e}$ and $M_{t\delta e}$ are the incremental tail load and pitching moment due to elevator (lb/deg, and in.-lb/deg, respectively), and $\Delta\delta_{ecm}$ is the incremental check maneuver elevator angle (deg),

$$\Delta\delta_{ecm} = -(1 + CBF)(\delta_{ebal} - \delta_{etrim}) \qquad (8.32)$$

where CBF the checkback factor $= 0$ for elevator returned to the original trim position and 0.50 for overcheck conditions, δ_{ebal} is the elevator angle required for steady-state maneuver at n_z (deg), and δ_{etrim} is the elevator angle at the beginning of the maneuver, $n_z = 1.0$ deg.

8.4.3 Maintaining Design Load Factors During Checked Maneuver

The problem with the simplified procedure defined in Eq. (8.30) is that the airplane load factor will exceed the design requirement. Consider the airplane lift balance equation at the design maneuver factor n_z as shown in Eq. (8.33):

$$n_z W = L_{wb} + L_{tnz=1} + \left(\frac{dL_t}{dn_z}\right)\Delta n_z \qquad (8.33)$$

where L_{wb} is the tail-off lift due to wing, body, and nacelles (lb); $L_{tnz=1}$ is the balancing tail load at $n_z = 1.0$ lb; and dL_t/dn_z is the balancing tail load per g (lb/g).

If the assumption is made that the checked maneuver elevator is applied instantaneously, then the airplane lift will be increased by an increment such that the desired design maneuver load factor will be exceeded. This may be seen by introducing the second term in Eq. (8.30) into Eq. (8.33) as shown in Eq. (8.34):

$$n_z W = L_{wb} + L_{tnz=1} + \left(\frac{dL_t}{dn_z}\right)\Delta n_z + L_{t\delta e}\Delta\delta_{ecm} \qquad (8.34)$$

If the total lift on the aircraft is to be maintained such that the load factor n_z does not exceed the design load factor, then an adjustment in load factor must be made in the balancing tail load increment so that the design requirement is not exceeded. This becomes an iterative process because the checked elevator $\Delta\delta_{ecm}$ is a function of the elevator required for the balance maneuver at the desired load factor n_z.

8.4.4 Time History Analysis Rational Methods

The equations of motion for elevator pitching maneuvers are developed in Sec. 2.4.1. Checked maneuver analyses may be accomplished by solving Eqs. (2.32) and (2.33) assuming a specific elevator input time history as is shown in Figs. 2.10 and 2.11.

Time history analyses are shown in Tables 2.8a–2.8b for an elevator checked maneuver using a sinusoidal elevator input as required by JAR 25.331(c).

Whatever shape of elevator input is used, the process requires iteration to obtain the required amplitude of elevator to produce the desired maximum load factor during the maneuver.

8.4.5 Comparison of Horizontal Tail Loads

Horizontal tail loads for checked maneuvers calculated using the methods discussed in this chapter are compared in Tables 8.3 and 8.4 with the time history analysis results for the JAR 25.331(c)(2) sinusoidal checkback elevator condition.

Table 8.3 Comparison of checked maneuver horizontal tail loads calculated
by different methods and criteria at forward center of gravity positions with
the following conditions: gross weight = maximum flight weight, $CG = 0.20$
%mac/100, altitude = sea level, V_e = 282 keas, and Mach = 0.426

		Methods/criteria		
	FAR 25.331 minimum[a]	Simplified methods		JAR 25.331(c)[d]
		Without rebalance[b]	With rebalance[c]	
n_{zcg}	2.50	2.66	2.50	2.50
θ, rad/s^2	−0.230	−0.378	−0.343	−0.360
$L_{tnz=1}$, lb	−6,362	−6,362	−6,362	−6,362
L_{tcm}, lb	25,381	40,476	36,068	38,200
M_{tcm}, 10^6 in.-lb	−0.651	−0.876	−0.811	−1.456

[a]Minimum requirement per FAR 25.331(c)(2) [see Eq. (2.38)].
[b]Using Eqs. (8.30) and (8.31). The elevator is checked back to neutral for this condition.
[c]Using Eqs. (8.30) and (8.31). The elevator is checked back to neutral for this condition.
[d]Time history analysis using sinusoidal elevator input discussed in Sec. 2.6. See Fig. 2.10.

The method of analysis discussed in Sec. 8.4.2 (the so-called rational approach) as used for aircraft certified before 1970 gave load factors at the airplane center of gravity that exceeded the design maneuver requirement for the airplane. This approach is conservative with respect to the tail load, as noted in Table 8.3. Tail loads are also shown where the checkback elevator was accomplished such that the required maneuver factor n_z, was not exceeded. This approach was used for transport aircraft certified after 1970 for tail load surveys to select the critical speed/altitude for a more in-depth study of check maneuver loads.

As noted in Table 8.3, the rational method gives horizontal tail loads that are greater than the JAR 25.331(c)(2) approach, but the resulting horizontal tail

Table 8.4 Comparison of checked maneuver horizontal tail loads calculated
by different methods and criteria at aft center of gravity positions with the
following conditions: gross weight = maximum flight weight, $CG = 0.30$
%mac/100, altitude = sea level, V_e = 282 keas, and Mach = 0.426

		Methods/criteria		
	FAR 25.331 minimum[a]	Simplified methods		JAR 25.331(c)[d]
		Without rebalance[b]	With rebalance[c]	
n_{zcg}	2.50	2.67	2.50	2.50
θ, rad/s^2	−0.230	−0.404	−0.363	−0.318
$L_{tnz=1}$, lb	−374	−374	−374	−374
L_{tcm}, lb	40,490	58,609	52,684	47,660
M_{tcm}, 10^6 in.-lb	−0.745	−1.595	−1.463	−1.676

[a]Minimum requirement per FAR 25.331(c)(2) [see Eq. (2.38)].
[b]Using Eqs. (8.30) and (8.31). The elevator is overchecked 50% for this condition.
[c]Using Eqs. (8.30) and (8.31). The elevator is overchecked 50% for this condition.
[d]Time history analysis using sinusodal elevator input discussed in Sec. 2.6. See Fig. 2.10.

pitching moment is less. This is due to the overcheck elevator used in the JAR procedure, which uses full 100% checkback of the elevator. It is not obvious that the JAR conditions are less critical than the method defined by Eqs. (8.30) and (8.31), but an in-depth stress analysis investigation is required.

As can be seen from Tables 8.3 and 8.4, horizontal tail check maneuver loads computed using the FAR 25.331(c)(2) minimum requirement are less than the loads computed from the rational method or the time history analysis results. This has been true on the 707, 727, 737, and 757 airplanes, but for the large aircraft such as the 747 series the tail loads calculated using minimum requirements have been greater than those calculated the rational methods.

It should be noted that the proposed revision to FAR/JAR 25.331 during the harmonization process removed the minimum requirement defined by Eqs. (2.37) and (2.38) and replaced it with the rational time history requirement.

8.4.6 Effect of Speedbrakes on Checked Maneuver Loads

Horizontal tail loads for checked maneuver conditions at positive design load factors may become more positive if the maneuver is accomplished with speedbrakes extended. This may also be true for power-on conditions, depending on the engine thrust line location with respect to the airplane center of gravity.

8.5 Vertical Gust Conditions

Vertical gust considerations are discussed in Chapter 5 with respect to the requirements and methods of analyses. The actual application of these criteria and methods to calculate horizontal tail loads due to vertical gusts is discussed in this section.

8.5.1 Gust Formula Approach

Traditionally horizontal tail loads have been determined for vertical gust using the gust formula approach as discussed in Sec. 5.5.1, where the airplane is assumed to translate vertically due to the gust encounter,

$$L_t = L_{tnz=1} + (1 - \epsilon_{\alpha w}) \Delta \alpha_w L_{t\alpha s} \qquad \text{(lb)} \qquad (8.35)$$

where

$$\Delta \alpha_w = 57.3 K_g U_{de}/(1.69 V_e) \qquad \text{(deg)} \qquad (5.14)$$

Since the airplane is assumed to not respond in pitch in the gust formula approach, only the downwash effect is considered in Eq. (8.35). In lieu of a rational analysis, the gust alleviation factor K_g is applied to the gust intensity for the horizontal tail. The horizontal tail incremental load defined by Eq. (5.15) is added to the initial balancing tail load corresponding to steady level flight as shown in Eq. (8.35).

8.5.2 One-DOF (Plunge Only) Analysis

The one-DOF (plunge only) analysis as developed in NACA Report 1206[3] is discussed in detail in Sec. 5.5.2. Horizontal tail gust loads obtained for this analysis are shown in Table 5.8 as a function of the gust penetration distance s. Incremental tail loads due to the gust are calculated using Eq. (5.23). The solution shown is

Table 8.5   Effect on vertical gust analysis—one-DOF solution using various definitions of the Küssner and Wagner functions

Condition code[a]	$\blacktriangle L_t$,[b] lb	Ratio
1[c]	26,015	1.0
2	28,737	1.105
3	26,769	1.029
4	24,330	0.935
Gust formula	26,462	1.017

[a]See Table 5.11 for the transient lift functions.
[b]$\blacktriangle L_{tg} = L_{\alpha s}(1 - \epsilon_{\alpha w})\blacktriangle\alpha_{wg}$, and $\blacktriangle\alpha_{wg}$ is obtained from the time history solution as a function of chords (s).
[c]The values shown for condition 1 are based on the analysis shown in Table 5.8.

based on the unsteady lift Küssner and Wagner functions as defined in NACA Report 1206,[3] identified in Table 5.11 as condition code 1.

The effect of using various definitions of the Küssner and Wagner functions is shown in Table 8.5. The reference source for the four conditions is given in Table 5.11.

8.5.3   Two-DOF Analysis

The two-DOF vertical gust analysis is discussed in detail in Sec. 5.5.3. Horizontal tail gust loads as obtained from this analysis are shown in Tables 5.9a–5.9c. Since gust penetration is considered, the example shows that peak load factors, pitching acceleration, and horizontal tail load occur at different times.

The gust formula data are also shown in Table 5.9a for comparison purposes. Since the regulations allow use of a rational analysis, the two-DOF analysis gives lower loads than the gust formula approach. This may be academic in determining horizontal tail gust loads because the horizontal tail structure is seldom critical for vertical gust conditions when compared with maneuver conditions for most jet transport aircraft.

8.5.4   Dynamic Load Considerations

Horizontal tail loads must be determined for both the discrete and continuous turbulence gust design criteria as discussed in Chapter 5 per the proposed requirements of FAR/JAR 25.341. Horizontal tail loads in general are not critical for vertical gust conditions as noted in the example shown in Fig. 8.9. Although only the discrete tuned gust condition is shown, continuous turbulence power spectral density (PSD) loads are of a similar magnitude and are not critical for structural design of the horizontal tail.

8.6   Unsymmetrical Load Conditions

In addition to the requirements for determining various symmetrical load conditions, the horizontal tail supporting structure is also designed by unsymmetrical load conditions. Known unsymmetrical conditions due to lateral gusts, rolling

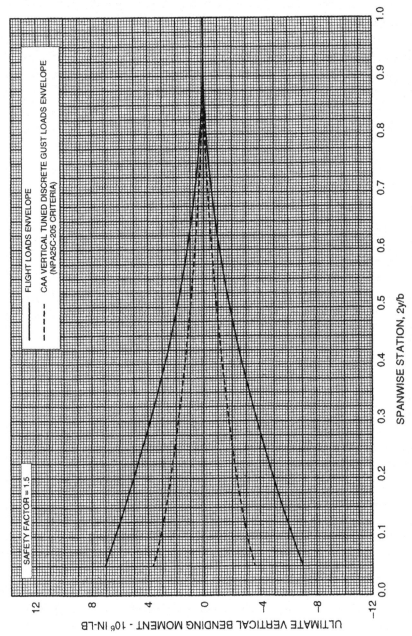

Fig. 8.9   Comparison of horizontal tail dynamic gust loads with the design maneuver envelope.

maneuvers, yawing maneuvers, buffet conditions occurring during stalls or high-speed turns, and elevator control system failures need to be considered during design.

In addition to these known conditions, unsymmetrical loads due to specified distributions of symmetrical conditions must be calculated.

Attention is given in this section to the regulatory requirements and the methods of analysis available that may be used to determine unsymmetrical loads.

8.6.1 Historical Perspective

In the prejet days, unsymmetrical loads on horizontal tails were computed per the requirements of CAR 4b.220(c) as follows[2]:

> Horizontal tail surfaces and supporting structure shall be designed for unsymmetrical loads arising from yawing and slipstream effects in combination with the prescribed flight conditions. In the absence of more rational analysis, the following assumptions may be made for airplanes that are conventional in regard to location of propellers, wings, tail surfaces, and fuselage shape: 1) 100% of the maximum loading from the symmetrical flight conditions acting on the surface on one side of the plane of symmetry and 80% of this load acting on the other side, and 2) where the design is not conventional (e.g., where the horizontal tail surfaces have appreciable dihedral or are supported by the vertical tail surfaces), the surfaces and supporting structures must be designed for combined vertical and horizontal surface loads resulting from the prescribed maneuvers.

For the 707-100/300 series aircraft, which were designed to CAR 4b,[2] the horizontal tail design was deemed conventional even though the tail dihedral was 7 deg. Thus the standard unsymmetrical loads were determined using the 100/80 distribution in lieu of a rational approach. These airplanes did have a significant unsymmetrical capability built into the structure due to the one-elevator inoperative failure condition that was a special design condition specified by the regulatory agency.

For the 727-100/200 airplanes, the horizontal tail design was not conventional (stabilizer on top of the fin), and special wind-tunnel tests were run to determine the unsymmetrical load conditions as we do in the more modern fleet. In addition, due to the low frequency of the so-called T-tail structural mode, special instrumentation was applied to the flight load survey airplane to determine loads during high-speed and low-speed buffet conditions. Again, as in earlier designs, a significant unsymmetrical capability was built into the structure because of the one-elevator inoperative failure condition.

8.6.2 Proposed FAR/JAR Requirements

The following are excerpts from the proposed revision to FAR/JAR 25.427 pertaining to horizontal tail unsymmetrical loads:

> (a) Horizontal tail surfaces and their supporting structure must be designed for unsymmetrical loads arising from yawing and sideslip effects, in combination with the prescribed flight condition.
>
> (b) In the absence of more rational data, the following apply:
>
> (1) For airplanes that are conventional in regard to location of propellers, wings, tail surfaces, and fuselage shape—

(i) 100% of the maximum loading from the symmetrical flight
condition may be assumed to act on the surface on one side of the
plane of symmetry; and

(ii) 80% of this loading may be assumed to act on the other side.

(2) For empennage arrangements where the horizontal tail surfaces
have appreciable dihedral or are supported by the vertical tail surfaces, the surfaces and supporting structure must be designed for the
combined vertical and horizontal surface loads resulting from each
prescribed flight load condition considered separately and for gust velocities specified in 25.341(a) acting at any orientation at right angles
to the flight path.

The addition of the around-the-clock gust (oblique gust) was agreed upon by the
two agencies during the 1992/1993 harmonization process. These types of gusts
are discussed in Sec. 5.10.

The problem the author has with the old and proposed revision to the regulations
is the interpretation of the words "appreciable dihedral." The question can be asked,
just how much dihedral is required such that the around-the-clock gust analysis
must be accomplished?

8.6.3   Arbitrary 100-80 Distribution

As noted earlier, in lieu of a rational analysis, unsymmetrical loads may be
determined from the symmetrical design flight conditions if the horizontal tail
structure is conventional. Unsymmetrical loads for arbitrary load distributions
may be calculated using either of the following two methods.

If the spanwise center of pressure and exposed surface airloads are calculated
for specific symmetrical load conditions, then the unsymmetrical moment about
the airplane centerline may be obtained from Eq. (8.36):

$$M_{\mathrm{CL}} = (1 - R) {\scriptstyle\blacktriangle} L_{t\,\exp} Y_{\mathrm{cp}} \qquad (8.36)$$

where $R = 0.80$ for the 100-80 distribution. The symmetrical and unsymmetrical
load conditions are defined in Fig. 8.10. An alternate method may be used when
the horizontal tail airloads at the side of the body as shown in Fig. 8.11 are used
to calculate the rolling moment about the airplane centerline.

The freestream moment at the side of the body may be calculated from Eq.
(8.37) using symmetrical design conditions at the side of the body as shown in
Fig. 8.11:

$$M_{\mathrm{FS}} = M_x \cos \Gamma_{\mathrm{sob}} + T_{lra} \sin \Gamma_{\mathrm{sob}} \qquad (8.37)$$

where M_x and T_{lra} are the bending moment and torsion referenced to the load
reference axis (in.-lb), and Γ_{sob} is the load reference angle at side of body (deg).

The rolling moment about the airplane centerline may now be calculated using
Eq. (8.38):

$$M_{\mathrm{CL}} = (1 - R)(M_{\mathrm{FS}} + V_z y_{\mathrm{sob}}) \qquad (8.38)$$

where y_{sob} is defined in Fig. 8.11. Unsymmetrical horizontal tail loads for a 100-80
distribution are shown in Table 8.6 using Eq. (8.38).

Table 8.6   Unsymmetrical horizontal tail loads based on 100-80 distribution per FAR/JAR 25.427(b)(1); rolling moments about the airplane centerline are calculated using Eq. (8.38) where $R = 0.80$ (see Fig. 8.11)

Condition type	Side[a]	Load, %	V_z, 10^3 lb	M_x, 10^6 in.-lb	T_{lra}, 10^6 in.-lb	M_{CL}, 10^6 in.-lb
Flaps down,	LHS	100	−22.30	−1.96	0.197	——
$n_z = 2.0$	RHS	80	−17.84	−1.57	0.158	−0.488
Balanced maneuver,	LHS	100	−15.86	−1.27	0.704	——
$n_z = 2.5$	RHS	80	−12.69	−1.02	0.563	−0.294
Positive checked	LHS	100	22.44	1.96	0.161	——
maneuver	RHS	80	17.95	1.57	0.129	0.506
Abrupt-up	LHS	100	−21.34	−1.87	0.298	——
elevator at V_A	RHS	80	−17.07	−1.50	0.238	−0.459

[a]LHS = left-hand side and RHS = right-hand side.

a)

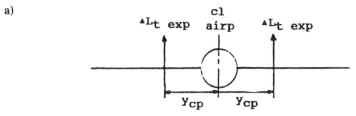

$^{▲}L_{Texp}$ = exposed suface airload, lbs

Y_{cp} = airload spanwise center of pressure, in.

b)

M_{cl} = [(100 − R)/100] $^{▲}L_t$ exp Y_{cp}

R = unsymmetrical load distribution factor

Fig. 8.10   Horizontal tail unsymmetrical loads per specified criteria using method 1: a) symmetrical loads and b) unsymmetrical loads.

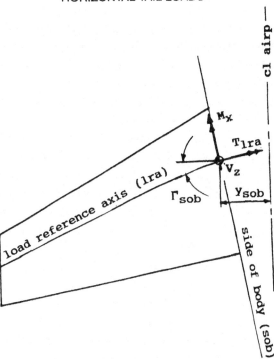

Fig. 8.11   Horizontal tail unsymmetrical loads per specified criteria using method 2.

The factor R shown in Eqs. (8.36) and (8.38) is the specified unsymmetrical load distribution required by the criteria used for design of the horizontal tail structure.

The arbitrary unsymmetrical load distributions required in military specifications vary from 100-50 distribution throughout the V-n diagram to 150-50 distribution of conditions flown at $C_{N\text{max}}$. At times to provide unsymmetrical capability for design, the military distributions may be used until further analyses are available.

In general, the 100-80 unsymmetrical conditions are not as critical for the horizontal tail center section and supporting structure design as other rational conditions, such as control system failure or buffet conditions.

8.6.4 Rolling Maneuver Conditions

As discussed in Sec. 3.5, rolling maneuvers are combined with the symmetrical flight maneuver loads at the prescribed load factors as specified in Table 3.2. In general, there are two conditions that may be considered for determining unsymmetrical horizontal tail loads due to rolling maneuvers.

The steady rolling condition may produce unsymmetrical horizontal tail loads due to roll velocity that are critical for the rolling moment about the airplane centerline. The horizontal tail rolling moments due to a steady roll, shown in Table 8.7, include static aeroelastic effects.

Except for configurations with large outboard masses at the horizontal tail tips, such as outboard fins, the loads due to rolling acceleration may be neglected as small. Both the roll initiation and termination should be considered.

Table 8.7   Unsymmetrical horizontal tail loads due to steady rolling maneuver

Altitude, ft	V_e, keas	Mach	V_t, ft/s	M_{CL},[a] 10^6 in.-lb
	V_A airspeeds, roll velocity = 1.25 rad/s[b]			
0	260	0.39	439	−0.70
20,000	260	0.58	602	−0.53
	V_C airspeeds, roll velocity = 1.25 rad/s[b]			
0	350	0.53	592	−0.93
20,000	350	0.78	810	−0.74
	V_D airspeeds, roll velocity = 0.42 rad/s[c]			
0	440	0.67	439	−0.39
9,900	440	0.80	602	−0.36

[a]Rolling moment about airplane centerline.
[b]Assumed design roll velocity.
[c]One-third roll rate at V_A.

8.6.5   Yawing Conditions Due to Gust and Maneuvers

Unsymmetrical horizontal tail loads must be determined for the yawing conditions due to rudder maneuvers (pilot induced or engine-out) as discussed in Secs. 4.2 and 4.3 and lateral gust conditions as discussed in Sec. 5.9.

The rolling moments due to sideslip and rudder may be obtained from data measured during wind-tunnel yaw tests using a strain-gauge balance at the horizontal stabilizer root. These data may be verified during flight test using the instrumentation required for stall buffet as discussed in Sec. 8.7.

8.6.6   Around-the-Clock Gust Conditions

The unsymmetrical horizontal tail loads due to around-the-clock gusts are not critical for the stabilizer center section supporting structure design but may be critical for the aft body design whereby the unsymmetrical horizontal tail loads are combined with the vertical tail loads acting on the aft fuselage structure.

The around-the-clock gust analysis is discussed in Sec. 5.10.

8.6.7   Unsymmetrical Elevator Conditions

Unsymmetrical elevator conditions due to control system failures must be investigated per FAR/JAR 25.671. Some of the typical failure conditions for which unsymmetrical horizontal tail loads are critical will be considered.

8.6.7.1   One-elevator inoperative conditions.   A one-elevator inoperative condition assumes one elevator is disconnected and is positioned by the centering springs or remains in a neutral position due to a single structural or hydraulic failure. The other elevator is commanded by the pilot to the blowdown limits or as required for the symmetrical maneuvers discussed in Chapter 2.

This type of maneuver may produce the critical design unsymmetrical condition for the horizontal tail center section and support structure.

8.6.7.2 Antisymmetrical elevator configurations. Aircraft certified after 1980 may have incorporated an antisymmetrical elevator capability to provide a limited override by the pilot resulting from control system jamming. This failure condition has a limited amount of $+/-$ elevator available due to the design of the breakout system between the left and right control column.

The special conditions applied to aircraft certified with this capability require consideration of both flaps down and up conditions at load factors from 0.50 to 1.50.

The resulting airload distribution is obtained for the unsymmetrical elevator conditions by addition of symmetrical loads for the flight maneuver under investigation to the antisymmetrical loads due to elevator as shown in Fig. 8.12.

As a tool to facilitate load surveys for horizontal tail loads due to unsymmetrical elevators, an investigation was made to determine the relationship between the symmetrical and antisymmetrical horizontal tail loads at the side of the body.

The antisymmetrical elevator factor may be defined as shown in Eq. (8.39):

$$K_{ae} = M_{xa}/M_{xs} \qquad (8.39)$$

where M_{xa} is the bending moment at the side of the body due to elevator, antisymmetrical load analysis (in.-lb), and M_{xs} is the bending moment at the side of the body due to elevator, symmetrical load analysis (in.-lb).

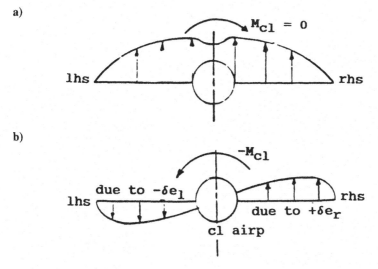

Fig. 8.12 Unsymmetrical horizontal tail loads due to split elevator failure condition: a) symmetrical distribution and b) antisymmetrical distribution.

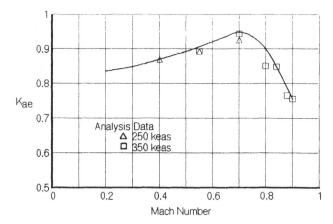

Fig. 8.13   Antisymmetrical elevator factor defined by Eq. (8.39).

The antisymmetrical elevator factor defined by Eq. (8.39) is shown in Fig. 8.13 as an example.

Using K_{ae} as shown in Fig. 8.13, unsymmetrical horizontal tail loads for an antisymmetrical elevator design failure condition are shown in Table 8.8 as an example. The use of the antisymmetrical factor simplified the work in determining the critical design elevator failure condition throughout the speed altitude conditions investigated. Since the symmetrical distribution of loads was necessary at each speed altitude condition studied, the antisymmetrical analysis was not run, thus reducing the time and effort to complete the load survey.

8.7   Stall Buffet Considerations

As commercial jet transport aircraft were developed, it became apparent that the horizontal tails of some aircraft configurations were subjected to a significant buffet occurring during stall entry.

Although FAR/JAR 25.201 stall demonstrations are done in straight level flight and in 30-deg banked turns, $n_z = 1.15$, the aircraft structure must be designed up to the design maneuver load factors at airplane C_{Nmax}, flaps up and down.

During stall flight testing the horizontal tail rocking mode increases with intensity as the stall entry progresses until the stall is completed as shown in Fig. 8.14. Maximum rolling moments measured during stall tests may be plotted as a function of exposed net horizontal tail load, airload plus inertia. These are then compared with the design envelope of the unsymmetrical flight conditions for evaluation.

FAR 25.305(e) requires the aircraft structure to be designed for buffeting that might occur in any likely operating condition up to V_D/M_D, including stall and probable inadvertent excursions beyond the boundaries of the buffet onset envelope. Usually special certification requirements applied to each new airplane program stipulate that buffet loads must be validated by flight tests.

8.8   High-Speed Buffet Considerations

On some aircraft configurations a similar phenomenon occurs during pull-ups or wind-up turns to design C_{Lmax} in the flaps-up configuration at high speeds.

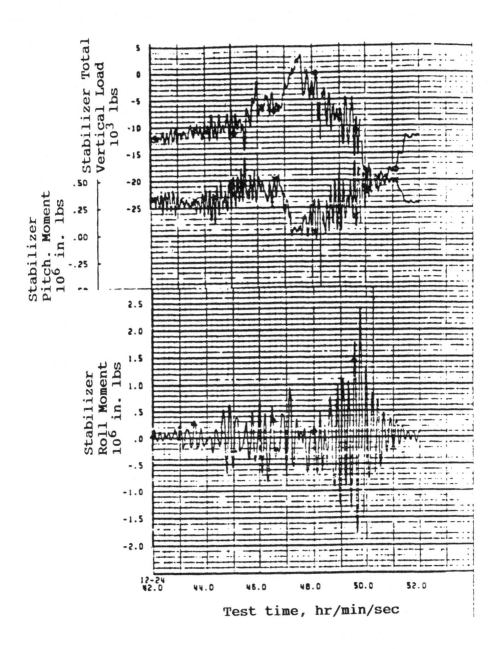

Fig. 8.14   Representative stall buffet loads on a horizontal tail.

Table 8.8   Unsymmetrical horizontal tail loads due to
antisymmetrical elevator design failure condition[a]

	Symmetrical loads[b]	Antisymmetrical loads[c]
α_s, deg	−6.47	——
δ_e, deg	−4.86	±13.0
n_{zcg}	1.50	——
K_{ae} (see Fig. 8.13)	——	0.850
Net loads at side of body		
V, 10^3 lb	−12.748	12.271
M_{FS}, 10^6 in.-lb	−1.357	1.387
$T_{pivot}10^6$ in.-lb	0.236	−0.445
M_{CL}, 10^6 in.-lb	0	3.867

[a]Loads are shown for the symmetrical maneuver condition at a positive load factor $n_z = 1.50$. The antisymmetrical condition is shown for a maximum split elevator available, with the following conditions: altitude = sea level, gross weight = maximum flight weight, $V_e = 210$ keas, CG = forward, and Mach = 0.32.

[b]Symmetrical loads at side of body include inertia and aeroelastic effects.

[c]Antisymmetical loads are calculated using Eq. (8.39).

During flight test of a commercial transport, high-speed buffet excited the horizontal tail rocking mode at Mach numbers below 0.70. At Mach numbers higher than 0.70 the tail rocking mode was not excited or was significantly diminished even though the airplane entered severe buffet.

A related concern based on flight test experience is the effect on the horizontal tail tip during high-speed pull-ups into buffet. The extreme tip of the horizontal tail may be excited in a torsion mode, like 15–18 Hz, such that high shear stresses may be created in the tail rear spar. This occurred outboard of 75% spanwise location, but again only for Mach numbers below 0.70. Structural analysis proved the adequacy of the shear stresses in the rear spar due to these buffet loads.

References

[1]Anon., "Part 25—Airworthiness Standards: Transport Category Airplanes," Federal Aviation Regulations, U.S. Dept. of Transportation, Jan. 1994.

[2]Anon., "Part 4b—Airplane Airworthiness: Transport Categories," Civil Air Regulations, Civil Aeronautics Board, Dec. 1953.

[3]Pratt, K. G., and Walker, W. G., "Revised Gust Load Formula and a Re-Evaluation of V-G Data Taken on Civil Transport Airplanes from 1933 to 1950," NACA Rept. 1206, Sept. 1953.

9
Vertical Tail Loads

Vertical tail loads affect the design of a significant part of the aircraft structure and thus require careful consideration of the various design requirements and resulting conditions. The structures affected by vertical tail loads are 1) the vertical tail and rudder, 2) the aft body structure, 3) the horizontal tail structure if the tail is mounted up the fin as on the DC-10 or on top of the fin like the 727 configuration, and 4) the fuselage center section (overwing) structure.

In general, the basic conditions that will determine the maximum loads for the vertical tail and related structure are 1) yawing maneuver conditions (pilot induced and engine out) and 2) lateral gust conditions.

Other conditions such as rolling maneuvers usually are not as critical for design of the vertical tail structure except possibly for structural configurations with horizontal tails mounted up on the fin.

The sign convention used for vertical tail load analyses is shown in Fig. 9.1.

9.1 Vertical Tail Loads for Yawing Maneuvers

The equations of motion for yawing maneuvers, defined by Eq. (4.14), are applicable to pilot-induced yawing maneuvers or engine-out conditions.

Using the generalized load parameter $\mathcal{L}$, vertical tail loads can be calculated from unit solution data as shown in Eq. (9.1):

$$\mathcal{L}_{\text{net}} = \mathcal{L}\alpha_v\alpha_v + \mathcal{L}_{\delta r}\delta_r + \mathcal{L}_{\text{in}}n_{yvt} \tag{9.1}$$

where $\mathcal{L}\alpha_v$ is the vertical tail load, moment, or torsion due to fin angle of attack = 1.0 deg; $\mathcal{L}_{\delta r}$ is the vertical tail load, moment, or torsion due to rudder = 1.0 deg; $\mathcal{L}_{\text{in}}$ is the vertical tail load, moment, or torsion due to inertia of $n_{yvt} = 1.0$; and δ_r is the rudder angle (deg).

The angle of attack of the fin α_v during the maneuver is defined by Eq. (9.2):

$$\alpha_v = -\beta + \blacktriangle\alpha_{vbb} \tag{9.2}$$

where β is the airplane sideslip angle (deg).

The relationship between vertical tail angle of attack and airplane sideslip angle is discussed in Sec. 9.8.

The change in vertical tail angle of attack due to aft body lateral bending is determined from Eq. (9.3), neglecting lateral loads due to inertia:

$$\blacktriangle\alpha_{vbb} = \left(\frac{d\alpha_v}{dL_{\alpha v}}\right)_{bb} L_v + \left(\frac{d\alpha_v}{dT_{\alpha v}}\right)_{bb} T_v \tag{9.3}$$

The net angle of attack of the fin may now be calculated from Eq. (9.4):

$$\alpha_v = (k_{bb})_{\alpha v}[-\beta + (k_{bb})_{\delta r}\delta_r] \tag{9.4}$$

143

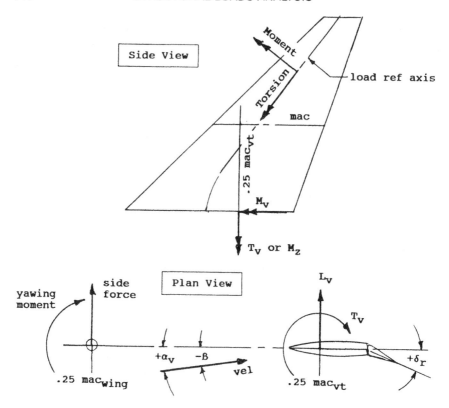

Fig. 9.1   Vertical tail loads sign convention.

The derivation of the static-elastic body-bending parameters, defined in Eqs. (9.5) and (9.6), is left up to the reader as an exercise. The effect on vertical tail angle of attack due to body bending is shown in Fig. 9.2;

$$(k_{bb})_{\alpha v} = 1 \bigg/ \left[1 - \left(\frac{d\alpha_v}{dL_{\alpha v}} \right)_{bb} L_{\alpha v} - \left(\frac{d\alpha_v}{dT_{\alpha v}} \right)_{bb} T_{\alpha v} \right] \qquad (9.5)$$

$$(k_{bb})_{\delta r} = \left(\frac{d\alpha_v}{dL_{\alpha v}} \right)_{bb} L_{\delta r} + \left(\frac{d\alpha_v}{dT_{\alpha v}} \right)_{bb} T_{\delta r} \qquad (9.6)$$

where $(d\alpha_v/dL_{\alpha v})_{bb}$ is the aeroelastic change in vertical tail angle of attack per unit vertical tail load (deg/lb), and $(d\alpha_v/dT_{\alpha v})_{bb}$ is the aeroelastic change in vertical tail angle of attack per unit vertical tail torsion (deg/in.lb).

9.2   Vertical Tail Loads for Rudder Maneuver Conditions

The rudder maneuver criteria as discussed in Sec. 4.2 of Chapter 4 lead to three conditions applicable to the fin, rudder, and supporting structure. These maneuvers, depicted schematically in Fig. 4.2, are defined in terms of vertical tail loads in this section.

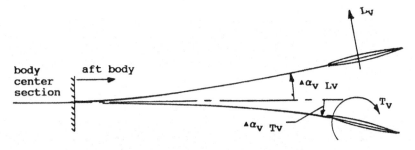

Fig. 9.2   Body lateral bending for static load analyses.

9.2.1   Abrupt Rudder Condition at Zero Yaw

With the airplane in unaccelerated flight at zero yaw, the vertical tail load condition occurring at the time of the application of maximum available rudder for the condition being investigated is identified as maneuver I in this book.

Maneuver I vertical tail loads may be calculated from the response parameters obtained from solution of the equations of motion, Eq. (4.14). Vertical tail loads, shown in Table 9.1, are calculated using Eq. (9.1) assuming a rigid body and neglecting inertia loads.

9.2.2   Maneuver I Vertical Tail Loads—Simplified Approach

Vertical tail loads may be determined for the maneuver I condition using the following simplified approach.

Defining the vertical tail loads due to maneuver I as the load due to rudder application times a load response parameter, and neglecting inertia loads, one can calculate net airload from Eq. (9.7):

$$\mathcal{L}_{vI} = k_1 \mathcal{L}_{\delta r} \delta_r \tag{9.7}$$

where $\mathcal{L}_{vI}$ is the vertical tail net airload, moment, or torsion for the maneuver I condition; $\mathcal{L}_{\delta r}$ is the vertical tail load, moment, or torsion due to rudder = 1.0 deg; δ_r is the rudder angle (deg); and k_1 is the airplane response factor and = (maneuver I load from time history analysis)/$(\mathcal{L}_{\delta r} \delta_r)$.

In general, a load survey is determined as shown in Table 9.1, assuming an airplane response parameter of 1.0 or using a conservative value of k_1, which may be verified by a time history analysis similar to Table 9.2.

9.2.3   Nonlinear Rudder Characteristics

Depending on the design characteristics of the rudder, the loads obtained for an abrupt maneuver condition may be very conservative when using a linear representation of the vertical tail load due to rudder for angles above 15 deg. This may be seen in Fig. 9.3.

An easy analysis technique that may be used to account for nonlinear effects at high rudder angles is to assume an effective rudder angle using linear analysis data that will produce the same airplane yawing moment as the nonlinear data. The simplified technique assumes that the reduction in rudder effectiveness applies to both the vertical tail lift and yawing moment due to rudder, which may not be true depending on the specific rudder under consideration.

Table 9.1 Vertical tail loads for rudder maneuvers using
simplified analysis based on Eqs. (9.7), (9.8), and (9.10)
(loads shown are limit)

Altitude, ft	0	0	0	0
V_e, keas	140	200	240	240
Mach	0.21	0.30	0.36	0.36
Flap position	All	All	0	0
$(k_{bb})_{\alpha v}$	1.0	1.0	1.0	0.962
$(\beta/\delta_r)_{ss}$	——	0.851	0.867	0.867
$k_{\delta r}$	0.85	1.0	1.0	1.0
Maneuver I loads				
δ_r, deg	26.8	11.0	7.2	——
α_v,[a] deg	0	0	0	——
L_v, lb	16,443	15,170	13,504	——
M_v, 10^6 in.-lb	2.178	1.986	1.748	——
T_v, 10^6 in.-lb	−1.335	−1.266	−1.156	——
Maneuver II loads				
δ_r, deg	——	11.0	7.2	——
Overyaw factor, deg	——	1.6	1.6	——
β_{oy}, deg	——	14.98	9.99	——
α_v,[a] deg	——	−14.98	−9.99	——
L_v, lb	——	−26,833	−26,039	——
M_v, 10^6 in.-lb	——	−3.403	−3.310	——
T_v, 10^6 in.-lb	——	−0.188	−0.154	——
Maneuver III loads				
δ_r, deg	——	0	0	0
β_{ss}, deg	——	9.36	6.24	6.24
α_v,[a] deg	——	−9.36	−6.24	−6.01
L_v, lb	——	−26,252	−26,714	−23,775
M_v, 10^6 in.-lb	——	−3.368	−3.161	−3.041
T_v, 10^6 in.-lb	——	+0.674	+0.626	+0.602

[a] α_v is calculated from Eq. (9.4).

The inclusion of nonlinear rudder characteristics may reduce high rudder angle loads such that the critical conditions are more likely to be in the linear region.

9.2.4 Rudder Maneuver Overyaw Conditions

With the rudder deflected as specified for the maneuver I condition, it is assumed that the airplane yaws to the resulting sideslip angle, usually called the "overyaw condition," or maneuver II in this book.

Maneuver II vertical tail loads may be calculated from the response parameters obtained from solution of the equations of motion, Eq. (4.14). Vertical tail loads, shown in Table 9.1, are calculated using Eq. (9.1) assuming a rigid body and neglecting inertia loads.

Table 9.2   Rudder maneuver time history, three-DOF analysis based on the
methodology shown in Eq. (4.14); the body is assumed rigid, and the rudder
input is linear at a rate of 40 deg/s[a]

Cond.: 2

Alt., ft = 0 V_e, keas = 240   Mach = 0.363

Gross weight, lb = 219,000   CG, % mac/100 = 0.352

Time, s	Rudder, deg	β, deg	n_{vt}	α_v, deg	L_v, lb	M_v, in.-lb 10^6	T_v, in.-lb 10^6
			Time history analysis ($\Delta t = 0.02$ s)				
0.20	7.20	0.102	0.307	−0.098	12,431	1.6340	−1.1160
0.40	7.20	0.469	0.271	−0.460	11,078	1.4583	−1.0832
0.60	7.20	1.071	0.221	−1.057	8,826	1.1664	−1.0281
0.80	7.20	1.862	0.159	−1.844	5,847	0.7809	−0.9552
1.00	7.20	2.789	0.091	−2.769	2,340	0.3270	−0.8691
1.20	7.20	3.799	0.019	−3.776	−1,489	−0.1681	−0.7750
1.40	7.20	4.834	−0.053	−4.812	−5,428	−0.6773	−0.6781
1.60	7.20	5.844	−0.120	−5.822	−9,279	−1.1749	−0.5833
1.80	7.20	6.781	−0.181	−6.760	−12,858	−1.6373	−0.4951
2.00	7.20	7.604	−0.232	−7.586	−16,011	−2.0444	−0.4173
2.20	7.20	8.281	−0.273	−8.266	−18,613	−2.3803	−0.3530
2.40	7.20	8.789	−0.300	−8.777	−20,577	−2.6335	−0.3044
2.60	7.20	9.115	−0.315	−9.107	−21,850	−2.7976	−0.2727
2.80	7.20	9.257	−0.317	−9.253	−22,421	−2.8708	−0.2583
3.00	7.20	9.220	−0.308	−9.219	−22,311	−2.8559	−0.2607
3.20	7.20	9.019	−0.287	−9.021	−21,573	−2.7600	−0.2786
3.40	7.20	8.675	−0.258	−8.680	−20,287	−2.5933	−0.3100
3.60	7.20	8.214	−0.222	−8.223	−18,555	−2.3691	−0.3524
3.80	7.20	7.669	−0.182	−7.679	−16,493	−2.1023	−0.4030
4.00	7.20	7.071	−0.139	−7.082	−14,225	−1.8090	−0.4587
			Time of maximum rudder—maneuver I				
0.18	7.20	0.079	0.310	−0.076	12,511	1.6445	−1.1179
			Time of maximum overyaw—maneuver II				
2.86	7.20	9.264	−0.316	−9.261	−22,458	−2.8753	−0.2573

[a] All loads shown are limit values, $SF = 1.0$. Airloads plus inertia loads are included.

9.2.5 Maneuver II Vertical Tail Loads—Simplified Approach

Vertical tail loads may be determined for the maneuver II condition using the
following simplified approach.

Defining the vertical tail airload due to maneuver II as the sum of the increments
due to overyaw (sideslip) and rudder, one can calculate loads from Eq. (9.8):

$$\mathcal{L}_{vII} = \mathcal{L}_\beta \beta_{oy} + \mathcal{L}_{\delta r} \delta_r \tag{9.8}$$

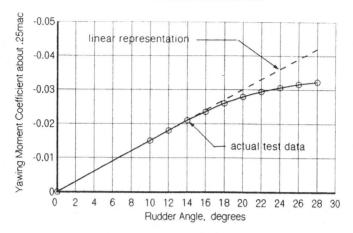

Fig. 9.3   Nonlinear rudder data.

where $\mathcal{L}_{vII}$ is the vertical tail net airload, moment, or torsion for the maneuver II condition; $\mathcal{L}_\beta$ is the vertical tail load, moment, or torsion due to sideslip = 1.0 deg; $\mathcal{L}_{\delta r}$ is the vertical tail load, moment, or torsion due to rudder = 1.0 deg; β_{oy} is the sideslip in the overyaw condition (deg); and δ_r is the rudder angle (deg).

The sideslip angle for the overyaw condition can be calculated from a time history analysis or obtained from flight test data as shown in Fig. 9.4. The overyaw angle then can be calculated from Eq. (9.9):

$$\beta_{oy} = F_{oy}\beta_{ss} \tag{9.9}$$

where the steady-state sideslip angle β_{ss} can be calculated from Eq. (4.3) or obtained from flight test data if available and F_{oy} is the overyaw factor.

Equation (9.8) may be modified to include body flexibility, but this is usually a sophistication that is unnecessary for this type of analysis.

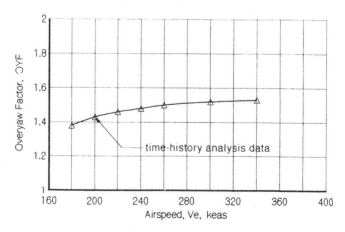

Fig. 9.4   Rudder maneuver overyaw factors.

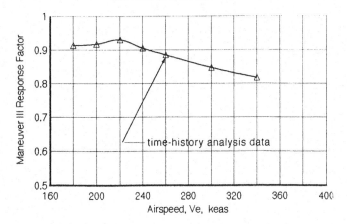

Fig. 9.5   Rudder maneuver III response factors.

9.2.6   Abrupt Rudder Checkback from a Steady Sideslip

With the airplane yawed to the static sideslip angle corresponding to the rudder deflection for maneuver I, it is assumed the rudder is abruptly returned to neutral. This condition is called maneuver III in this book.

Vertical tail loads for the maneuver III condition may be calculated by superposition of the tail loads during a steady sideslip with the increment in tail load due to rudder as obtained from the maneuver I condition, but with the opposite sign.

Since the vertical tail loads for the maneuver III condition are a function of the airplane steady sideslip, loads may be calculated directly from Eq. (9.10):

$$\mathcal{L}_{v\text{III}} = k_{\text{res}} \mathcal{L}_{\beta} \beta_{\text{ss}} \tag{9.10}$$

where $\mathcal{L}_{v\text{III}}$ is the vertical tail net airload, moment, or torsion for the maneuver III condition; $\mathcal{L}_{\beta}$ is the vertical tail load, moment, or torsion due to sideslip $= 1.0\,\text{deg}$; β_{ss} is the steady sideslip angle (deg); and k_{res} is the airplane response parameter.

The steady sideslip angle β_{ss} may be determined from Eq. (4.3) or obtained from flight test data. The airplane response parameter k_{res} may be determined from time history solutions to account for the reduction in sideslip angle as the rudder is returned to neutral. An example response parameter is shown in Fig. 9.5.

Equation (9.10) may be modified to include the body flexibility correction factor $(k_{\text{bb}})_{\alpha v}$ defined by Eq. (9.5),

$$\mathcal{L}_{v\text{III}} = k_{\text{res}} (k_{\text{bb}})_{\alpha v} \mathcal{L}_{\beta} \beta_{\text{ss}} \tag{9.11}$$

Vertical tail loads for rudder maneuver III conditions are shown in Table 9.1 calculated using both Eqs. (9.10) and (9.11).

The sophistication of the analysis will depend on the criticality of maneuver III loads, when compared with loads resulting from lateral gust conditions.

9.2.7   Comparison of Simplified Approach with Time History Loads

A comparison of the vertical tail loads obtained from the simplified approach using Eqs. (9.7), (9.8), and (9.10) with the results of the three-DOF time history analysis are shown in Table 9.3.

Table 9.3 Comparison of vertical tail loads[a] from simplified analysis using Eqs. (9.7), (9.8), and (9.10) with three-DOF time history results given the following condition parameters: altitude = sea level, gross weight = 219,000 lb, V_e = 240 keas, and CG = 0.352 % mac/100

	δ_r, deg	β, deg	n_{yt}	L_v, lb	M_v, 10^6 in.-lb	T_v, 10^6 in.-lb
Maneuver I conditions						
Time history	7.2	0.079	0.31	12,511	1.645	−1.118
Simplified	7.2	0	Neglect	13,504	1.748	−1.156
Maneuver II conditions						
Time history	7.2	9.264	−0.316	−22,458	−2.875	−0.257
Simplified	7.2	9.79	Neglect	−26,039	−3.310	−0.154
Maneuver III conditions						
Time history	0	6.041	Neglect	−23,917	−3.059	0.606
Simplified	0	6.12	Neglect	−24,714	−3.161	0.626

[a]All loads shown are limit. Airloads and inertia are included in time history data. The body is assumed to be rigid.

9.2.8 Concluding Remarks About Rudder Maneuver Conditions

As noted in Fig. 9.6, rudder maneuvers I and II form part of the design load envelope for the vertical tail and aft body. The following comments are given based on the experience of the author.

1) Rudder maneuver I vertical tail loads may be reduced using a time history analysis, particularly if flight test response parameters are available.

2) Rudder maneuver II vertical tail loads are dependent on the maximum overyaw during the condition. Time history results will allow determination of the maximum sideslip directly. This allows validation of the overyaw factor used in the load survey.

3) If maneuver III loads are less than lateral gust conditions, the loads calculated using the simplified approach are adequate.

For aircraft where the maneuver III loads exceed the lateral gust conditions, the maneuver III condition may be calculated directly from superposition of the maneuver I (with opposite signs) and the steady sideslip loads at the condition being investigated.

9.3 Vertical Tail Loads Engine-Out Conditions

The engine-out (EO) criteria as discussed in Sec. 4.3 lead to three conditions that are applicable to the fin, rudder, and supporting structure. These three conditions may be defined as follows: EO maneuver I: the vertical tail loads occurring at the time of maximum sideslip in the engine-out condition with zero rudder (no pilot response), EO maneuver II: the vertical tail loads occurring at the time of maximum yaw rate or at the time of corrective rudder application but not sooner than $t = 2.0$ s, and EO maneuver III: the vertical tail loads at zero sideslip with the rudder as required.

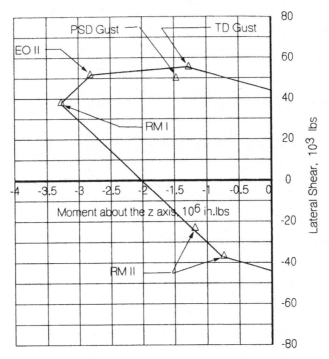

Fig. 9.6   Vertical tail load design envelope at root; loads shown include airload plus inertia. RM I—rudder maneuver I; RM II—rudder maneuver II; EO II—engine-out maneuver II; PSD Gust—continuous turbulence analysis; TD Gust—tuned discrete gust.

9.3.1   Thrust Decay for Engine-Out Conditions

Thrust decay becomes important in determining the resulting vertical tail loads due to engine-out conditions. Consideration must be given to the two types of failures discussed in Sec. 4.3.

Simulation of the engine-out condition due to fuel flow interruption, shown in Fig. 4.6, is considered a limit load condition and is represented in this figure as a linear variation of thrust for convenience of analysis.

Simulation of the engine-out condition due to mechanical failure may be assumed to act nearly instantaneously; hence T/T_0 shown in Fig. 4.5 may be assumed to be zero in 0.02–0.10 s for the time history analysis. This type of failure is considered an ultimate load condition.

The yawing moment produced by the failed engine would be greater than the value used in the example for the fuel flow interruption, which will windmill after the failure. The extreme condition would be a failure whereby the engine has stopped due to jamming and thus would produce significantly more drag than the windmilling case.

9.3.2   Load Analysis Using a Simplified Approach

Engine-out maneuver loads may be calculated in a similar manner as for rudder maneuvers using a simplified approach. The assumption is that the maneuver is accomplished in a wings level attitude and inertia loads are neglected.

Vertical tail loads for the engine-out maneuver I condition may be calculated using a modification of Eq. (9.11):

$$\mathcal{L}_{eo1} = k_{res}(k_{bb})_{\alpha v}\mathcal{L}_\beta \beta_{oy} \tag{9.12}$$

$$\mathcal{L}_{eo1} = k_{res}(k_{bb})_{\alpha v}\mathcal{L}_\beta F_{oy}\beta_{eo\ ss} \tag{9.13}$$

where $\mathcal{L}_{eo1}$ is the vertical tail net airload, moment, or torsion for the engine-out maneuver I condition; and $\beta_{eo\ ss}$ is the engine-out steady sideslip angle (deg).

The overyaw factor F_{oy} may be assumed the same as for the rudder maneuver II condition, until checked with a time history analysis.

The steady sideslip angle due to engine out, $\beta_{eo\ ss}$, may be calculated from Eq. (4.8) or obtained from flight test data. An approximation of the steady sideslip angle for an engine-out condition may be obtained using a simplification of Eq. (4.8):

$$\beta_{eo\ ss} = -Cn_{eo}/C_{n\beta} \tag{9.14}$$

The engine-out yawing coefficient Cn_{eo} may be determined from Eq. (4.6).

Vertical tail loads for the engine-out maneuver II condition may be calculated using a modification of Eq. (9.8):

$$\mathcal{L}_{eo2} = \mathcal{L}_\beta \beta_{eo} - \mathcal{L}_{\delta r}\delta_r \tag{9.15}$$

where $\mathcal{L}_{eo2}$ is the vertical tail net airload, moment, or torsion for the engine-out maneuver II condition; and β_{eo} is the sideslip angle associated with the time of maximum yaw rate or 2.0 s, whichever is less.

It should be noted that for the engine-out maneuver II condition the load due to rudder is additive to the sideslip increment, whereas in the rudder maneuver II condition the loads due to rudder and sideslip are opposite in sign. The engine-out maneuver II condition is essentially a bending moment case, whereas the rudder maneuver II condition is basically a fin torsion case.

Vertical tail loads for the engine-out maneuver III condition may be calculated using a modification of Eq. (9.7):

$$\mathcal{L}_{eo3} = (k_{bb})_{\delta r}\mathcal{L}_{\delta r}\delta_r \tag{9.16}$$

In the practical sense this condition must be less than the rudder maneuver I conditions, which are based on the maximum available rudder at all airspeeds up to V_D, except at V_{MC}, which usually requires maximum available rudder. If inertia loads are included in the rudder maneuver I analyses, then the engine-out conditions may be more critical for airspeeds where the maximum available rudder is required to control the engine-out sideslip. Inertia loads for the engine-out maneuver III are low because the condition is essentially a balanced maneuver in the lateral sense, and the translational load factors are small.

9.3.3 Typical Load Survey for Engine-Out Conditions

Typical load analysis surveys are shown in Table 9.4 for the engine-out maneuver conditions discussed in this section.

A conservative overyaw factor is assumed for the engine-out maneuver I conditions. For very low airspeed conditions at V_{MC}, the airplane minimum control

Table 9.4 Vertical tail loads for engine-out maneuvers
using simplified analysis based on Eqs. (9.13), (9.15), and (9.16)[a]

	Assumes rigid body		Assumes flexible body
Altitude, ft	0	0	0
V_e, keas	180	240	240
Mach	0.27	0.36	0.36
Flap position	0	0	0
$(k_{bb})_{\alpha v}$	1.0	1.0	0.962
Cn_{β}/deg	0.00326	0.00316	0.00316
$Cn_{\delta r}$/deg	−0.00275	−0.00274	−0.00274
Cn_{eo}	+0.02238	+0.01188	+0.01188
Engine-out maneuver I loads			
δ_r, deg	0	0	0
Overyaw factor	1.6	1.6	1.6
β, deg	−10.98	−6.02	−6.02
α_v, deg	10.98	6.02	5.79
L_v, lb	25,180	23,815	22,910
M_v, 10^6 in.-lb	3.236	3.046	2.930
T_v, 10^6 in.-lb	−0.650	−0.603	−0.580
Engine-out maneuver II loads			
δ_r, deg	8.14	4.34	——
k_{II}, deg	0.80	0.80	——
β_{II}, deg	−5.49	−3.01	——
α_v, deg	5.49	3.01	——
L_v, lb	21,911	20,039	——
M_v, 10^6 in.-lb	2.843	2.576	——
T_v, 10^6 in.-lb	−1.095	−0.998	——
Engine-out maneuver III loads			
δ_r, deg	8.14	4.34	——
α_v, deg	0	0	——
L_v, lb	9,321	8,132	——
M_v, 10^6 in.-lb	1.225	1.053	——
T_v, 10^6 in.-lb	−0.770	−0.696	——

[a]Loads shown are limit.

speed with an engine out, the resulting load on the vertical tail may be limited by the vertical tail maximum lift capability as shown in Table 9.4.

For the engine-out maneuver II conditions shown in Table 9.4, the assumption is made that the pilot initiates corrective rudder at $t = 2.0$ s. A further assumption may be made that, after the application of corrective rudder, the maximum sideslip angle obtained does not exceed the steady sideslip angle defined by Eq. (9.14). The resulting tail loads, although conservative, may be adequate for a load analysis survey as shown in Table 9.4 to ascertain the critical engine-out flight condition for this maneuver.

A load survey is shown in Table 9.4 for the engine-out maneuver III condition where the airplane is assumed to be balanced by the rudder required for zero sideslip. As noted, the vertical tail loads are low when compared with the rudder maneuver I conditions shown in Table 9.1.

9.3.4 Load Analysis Using a Time History Method

Vertical tail loads for the three-DOF engine-out analyses discussed in Sec. 4.4.2 are shown in Tables 9.5 and 9.6. Body-bending effects were neglected in these analyses, but inertia loads are included. The thrust decay used for these analyses is conservative when compared with the fuel flow interruption curve shown in Fig. 4.5.

Engine-out maneuver I maximum loads will occur at the time of maximum overyaw, as shown in the example analysis in Table 9.5. The overyaw factor of $\beta_{oy}/\beta_{ss} = 1.39$ is less than the value of 1.6 used in the simplified method shown in Table 9.4. The steady-state sideslip angle β_{ss} is calculated from Eq. (9.14).

Engine-out maneuver II maximum loads will occur at the time of maximum overyaw, as shown in the example analysis in Table 9.6. Because the time at maximum yaw rate is less than 2.0, the corrective rudder is applied at 2.0 s.

Because the regulations do not specify the amount of corrective rudder, the application of maximum available rudder may be used unless the condition becomes overly critical when compared with other vertical tail loads, such as rudder maneuver or lateral gust conditions.

9.4 Vertical Tail Loads Using the Gust Formula Approach

The proposed lateral gust criteria defined in FAR/JAR 25.341, "Gust and Turbulence Loads," eliminates the gust load formula requirements of FAR 25.351(b) and replaces them with the more rational discrete and continuous turbulence analyses.

Because the lateral gust load formula is interesting from a historical perspective, the author still believes that there is merit in completing a load survey of gust loads using the simplified approach.

9.4.1 Lateral Gust Load Formula (Historical Perspective)

The original lateral gust criteria in CAR 4b,[1] before March 1956, stipulated an average pressure distribution on the vertical tail using the gust velocities shown in Table 5.1. A loading equation, in terms of pound-force per square inch, was derived similar to the vertical gust load factor formula shown in Eq. (5.1).

After incorporation of Amendment 4b-3 of CAR 4b,[1] the lateral gust formula was introduced based on the airplane mass ratio in yaw as discussed in Sec. 5.9.1. In the absence of a rational analysis of the airplane response to a lateral discrete gust, the gust load formula previously defined in FAR 25.351(b) (before the proposed revision by the FAA/JAR harmonization process) has been used to determine the gust airload acting on the vertical tail.

The lateral gust load formula, assuming a rigid body, may be represented by Eq. (9.17):

$$L_v = K_g U_{de} V_e C_{Lav} S_v / 498 \qquad (9.17)$$

where K_g is the gust alleviation factor,

$$K_g = 0.88 \mu_v / (5.3 + \mu_v) \qquad (5.5)$$

Table 9.5 Engine-out maneuver time history analysis for maneuver I (zero
rudder) based on the methodology shown in Eq. (4.14); the body is assumed
rigid, and the rudder input is zero for the maneuver I condition[a]

Cond.: 1
Alt., ft $= 0$ V_e, keas $= 240$ Mach $= 0.363$
SIGF $= 1$ Gross weight, lb $= 219{,}000$ CG, % mac/100 $= 0.352$
$Cn_{eomax} = 0.01188$ Thrust decay, s $= 1.00$

Time, s	Rudder, deg	β, deg	n_{vt}	α_v, deg	L_v, lb	M_v, in.-lb 10^6	T_v, in.-lb 10^6
		Time history analysis ($\Delta t = 0.02$ s)					
0.20	0.00	−0.007	−0.029	0.007	93	0.0096	−0.0035
0.40	0.00	−0.050	−0.055	0.049	314	0.0360	−0.0102
0.60	0.00	−0.156	−0.075	0.153	772	0.0931	−0.0226
0.80	0.00	−0.348	−0.088	0.343	1,555	0.1923	−0.0430
1.00	0.00	−0.645	−0.093	0.637	2,730	0.3422	−0.0730
1.20	0.00	−1.048	−0.061	1.038	4,247	0.5386	−0.1101
1.40	0.00	−1.535	−0.025	1.524	6,088	0.7768	−0.1552
1.60	0.00	−2.076	0.014	2.064	8,139	1.0421	−0.2056
1.80	0.00	−2.642	0.053	2.628	10,287	1.3198	−0.2584
2.00	0.00	−3.201	0.091	3.188	12,419	1.5953	−0.3109
2.20	0.00	−3.728	0.126	3.716	14,432	1.8554	−0.3605
2.40	0.00	−4.200	0.156	4.189	16,236	2.0884	−0.4050
2.60	0.00	−4.596	0.180	4.587	17,758	2.2850	−0.4426
2.80	0.00	−4.904	0.197	4.896	18,944	2.4379	−0.4720
3.00	0.00	−5.113	0.207	5.108	19,760	2.5430	−0.4922
3.20	0.00	−5.222	0.210	5.218	20,191	2.5984	−0.5030
3.40	0.00	−5.231	0.207	5.230	20,243	2.6048	−0.5045
3.60	0.00	−5.148	0.198	5.148	19,942	2.5656	−0.4972
3.80	0.00	−4.983	0.183	4.985	19,326	2.4857	−0.4822
4.00	0.00	−4.749	0.164	4.752	18,448	2.3720	−0.4607
		Time of maximum overyaw					
3.32	0.00	−5.239	0.209	5.237	20,266	2.6080	−0.5050
		Time of maximum yaw rate					
1.86	0.00	−2.811	0.065	2.798	10,933	1.4033	−0.2743

[a] All loads shown are limit values, $SF = 1.0$. Airloads plus inertia loads are included.

where μ_v is the lateral mass ratio,

$$\mu_v = 2I_z / \left(\rho c_v Cl_{\alpha v} S_v l_v^2 \right) \tag{9.18}$$

and where U_{de} is the derived gust velocity (fps eas), V_e is the airplane equivalent
airspeed (keas), $C_{L\alpha v}$ is the lift curve slope of the vertical tail (per radian), S_v is
the vertical tail reference area (ft²), I_z is the moment of inertia in yaw (slug ft²),
ρ is the density of air at altitude (slug/ft³), c_v is the mean geometric chord of the

Table 9.6 Engine-out maneuver time history analysis for maneuver II based on the methodology shown in Eq. (4.14); the body is assumed rigid, and the corrective rudder input is as required for zero sideslip[a]

Cond.: 1

Alt., ft = 0		V_e, keas = 240			Mach = 0.363	
SIGF = 1		Gross weight, lb = 219,000			CG, % mac/100 = 0.352	
		$Cn_{eomax} = 0.01188$	Thrust decay, s = 1.00			

Time, s	Rudder, deg	β, deg	n_{vt}	α_v, deg	L_v, lb	M_v, in.-lb 10^6	T_v, in.-lb 10^6

Time history analysis ($\Delta t = 0.02$ s)							
0.20	0.00	−0.007	−0.029	0.007	93	0.0096	−0.0035
0.40	0.00	−0.050	−0.055	0.049	314	0.0360	−0.0102
0.60	0.00	−0.156	−0.075	0.153	772	0.0931	−0.0226
0.80	0.00	−0.348	−0.088	0.343	1,555	0.1923	−0.0430
1.00	0.00	−0.645	−0.093	0.637	2,730	0.3422	−0.0730
1.20	0.00	−1.048	−0.061	1.038	4,247	0.5386	−0.1101
1.40	0.00	−1.535	−0.025	1.524	6,088	0.7768	−0.1552
1.60	0.00	−2.076	0.014	2.064	8,139	1.0421	−0.2056
1.80	0.00	−2.642	0.053	2.628	10,287	1.3198	−0.2584
2.00	0.00	−3.201	0.091	3.188	12,419	1.5953	−0.3109
2.20	4.20	−3.644	0.302	3.634	21,593	2.7968	−1.0093
2.40	4.20	−3.873	0.309	3.868	22,502	2.9137	−1.0321
2.60	4.20	−3.895	0.302	3.894	22,622	2.9284	−1.0354
2.80	4.20	−3.724	0.283	3.727	22,001	2.8476	−1.0205
3.00	4.20	−3.380	0.252	3.387	20,724	2.6819	−0.9893
3.20	4.20	−2.894	0.214	2.904	18,897	2.4454	−0.9446
3.40	4.20	−2.299	0.169	2.311	16,651	2.1547	−0.8895
3.60	4.20	−1.632	0.121	1.645	14,123	1.8277	−0.8274
3.80	4.20	−0.931	0.071	0.944	11,457	1.4830	−0.7619
4.00	4.20	−0.231	0.024	0.245	8,793	1.1387	−0.6963

Time of maximum tail load							
2.54	4.20	−3.910	0.306	3.908	22,666	2.9344	−1.0364

[a]All loads shown are limit values, $SF = 1.0$. Airloads plus inertia loads are included.

vertical tail (ft), and l_v is the distance from airplane center of gravity to vertical tail center of lift (ft).

If vertical tail unit load parameters are available, it may be convenient to substitute Eq. (9.19) into Eq. (9.17):

$$C_{L\alpha v} S_v = 57.3 L_{\alpha v}/q \qquad (\text{ft}^2/\text{rad}) \qquad (9.19)$$

where $L_{\alpha v}$ is the vertical tail load due to fin angle of attack (lb/deg), and q is the dynamic pressure (lb/ft²).

The gust alleviation factor K_g as defined in Eq. (5.5) has the same form as that used for the vertical gust load factor formula, except the mass ratio is calculated using lateral parameters as noted in Eq. (9.18).

Modification of the gust load formula to include body lateral bending effects has been included in some historical analyses. The effects of aft body bending due to vertical load and torsion are shown in Fig. 9.2. The lateral gust load formula then is modified as shown in Eq. (9.20) by the insertion of the body-bending factor $(k_{bb})_{\alpha v}$ defined by Eq. (9.5):

$$L_v = (k_{bb})_{\alpha v} K_g U_{de} V_e C_{L\alpha v} S_v / 498 \qquad (9.20)$$

A comparison of the lateral gust loads computed using the gust load formula modified to included static body-bending effects is shown in Table 9.7.

Table 9.7   Vertical tail loads[a] calculated using the gust load formula
with Eq. (9.17) for the rigid-body analysis and Eq. (9.20) for the
flexible-body analysis; body flexibility parameters are calculated from
Eq. (9.5), and the condition parameters are $S_y = 370$ ft² and $(k_{bb})_{\dot\alpha v}$
$= 1/[1 + 0.009\,(10^{-3})\,L_{\dot\alpha v} - 0.036\,(10^{-6})T_{\dot\alpha v}$

Alt., ft	0	16,000	20,000	23,200
V_e, keas	350.0	341.1	337.9	335.0
Mach	0.529	0.700	0.754	0.800
σ	1.0000	0.6090	0.5328	0.4773
Weight, lb	239,000	237,800	237,200	236,600
CG, % mac/100	0.257	0.258	0.259	0.26
I_z, E-6 slug ft²	8.937	8.935	8.933	8.931
l_v, ft	61.96	61.94	61.93	61.91
$L_{\alpha v}$, lb/deg	8,060	8,162	8,264	8,338
$M_{\alpha v}$, E-6 in.-lb/deg	1.020	1.039	1.053	1.064
$T_{\alpha v}$, E-6 in.-lb/deg	−0.192	−0.206	−0.207	−0.208
$(k_{bb})_{\alpha v}$	0.926	0.925	0.924	0.924
μ_v	116.5	179.6	199.2	216.8
K_g	0.842	0.855	0.857	0.859
U_{de}, fps eas	50.0	50.0	50.0	47.3
Rigid-body analysis				
α_v, deg	4.079	4.250	4.300	4.113
L_v, lb ult.	49,315	52,033	53,307	51,440
M_v, E-6 in.-lb ult.	6.241	6.624	6.792	6.564
T_v, E-6 in.-lb ult.	−1.175	−1.313	−1.335	−1.283
Flexible-body analysis				
α_v, deg	3.777	3.931	3.974	3.800
L_v, lb ult.	45,666	48,131	49,256	47,531
M_v, E-6 in.-lb ult.	5.779	6.127	6.276	6.065
T_v, E-6 in.-lb ult.	−1.088	−1.215	−1.234	−1.186

[a]Ultimate loads are shown, $SF = 1.5$. Inertia load relief is neglected for this example.

**Table 9.8   Comparison of vertical tail root bending moments for loads
calculated using the lateral gust formula with PSD dynamic analysis results
based on the minimum gust values shown in Fig. 5.4**

				Gust formula		PSD	
				U_{de} or U_{ds},		U_σ,	
	Altitude,	V_e,		ft/s	Mv_{root},[a]	ft/s	Mv_{root},[a]
Speeds	ft	keas	Mach	eas	10^6 in.-lb	tas	10^6 in.-lb
V_B (original design gust velocities)							
	0	290	0.44	66.0	6.63	99.0	7.22
	20,000	283	0.63	66.0	7.02	99.0	6.01
V_C (original design gust velocities)							
	0	350	0.53	50.0	5.78	75.0	7.19
	20,000	338	0.75	50.0	6.28	75.0	6.97
	27,100	331	0.86	44.1	5.79	69.7	6.75
V_B (1.32 × proposed gust velocities per FAR/JAR 25.341)							
	0	290	0.44	59.22	5.95	99.0	7.22
	20,000	283	0.63	49.00	5.21	99.0	6.01
V_C (proposed gust velocities per FAR/JAR 25.341)							
	0	350	0.53	44.86	5.19	75.0	7.19
	20,000	338	0.75	37.12	4.66	75.0	6.97
	27,100	331	0.86	35.10	4.61	69.7	6.75

[a] Mv_{root} = ultimate vertical tail root bending moment ($SF = 1.5$).

9.4.2   Comments on Loads Survey Using the Gust Load Formula

The use of the gust load formula to determine criticality for dynamic load conditions has merit, providing that the gust velocity requirements are equivalent.

Consider the relationship between the old requirement of a 50 fps eas and the PSD levels used for commercial transport aircraft certified before 1990. Using the lateral gust load formula, Eq. (9.17), the critical condition at V_C would be at 20,000 ft, whereas the critical design PSD condition may be at sea level. This can be seen in the comparisons of vertical tail loads in Table 9.8.

If the gust velocities of the proposed FAR/JAR 25.341 criteria are used as obtained from Fig. 5.2 and Eq. (5.8), the gust formula loads would show that the critical condition is at sea level as the PSD gust conditions are. This again may be seen in the analysis shown in Table 9.8.

9.5   Lateral Gust Dynamic Analyses

Lateral gust dynamic analyses must be calculated for both discrete and continuous turbulence requirements. Consideration must be given for the use of a load alleviation system such as a yaw damper in determining the criticality of the resulting vertical tail and aft body loads. Horizontal tail loads may also be affected if the tail is mounted on top of the fin or in a midlocation such as on the Lockheed L-1011.

Comparison of various design load conditions, including dynamic lateral gust conditions, are shown in Table 9.8. These loads are also summarized in Fig. 9.6 as a means of comparing one load set with another. Although this figure is plotted as shear vs torsion, a similar plot of bending moment vs torsion would provide the same use.

9.6 Definition of Vertical Tail for Structural Analysis

The definition of the vertical tail for structural analysis will vary with the configuration, but it has been common practice to define the aerodynamic lifting surface as shown in Fig. 9.7. In essence, a portion of the aft body is shown as being part of the vertical tail surface from an aerodynamic viewpoint if the horizontal tail is mounted on the aft fuselage below the vertical tail. Usually the bottom of the vertical tail will be defined at the plane consistent with the horizontal tail at the side-of-body at a zero angle with respect to the body reference axis.

The 727 and BAC 111 aircraft do not fall into this category as the horizontal tail is mounted on top of the vertical tail.

9.7 Lateral Bending-Body Flexibility Parameters

Lateral bending-body flexibility parameters, as shown in Table 9.9, are applicable to both yawing maneuver and lateral gust static loads analyses.

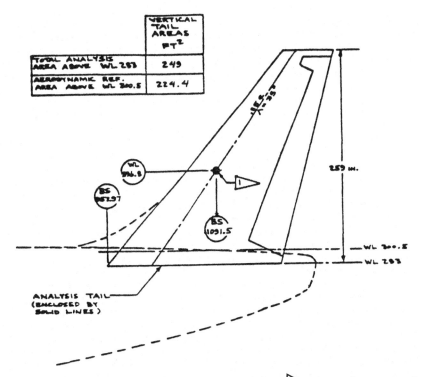

Fig. 9.7 Definition of vertical tail effective area. Notes: ▷ center of pressure of total load due to angle of attack for rigid tail; WL 283 in the horizontal tail plane at zero stabilizer position.

Table 9.9   Lateral bending body flexibility
parameters used for analyses as defined in
in Eqs. (9.5) and (9.6) (see Fig. 9.2) for the
airplane used to demonstrate the analytical
method of this chapter

$(10^{-3})\ d\alpha_v/dL_{av}$, deg/lb	-0.0090
$(10^{-6})\ d\alpha_v/dT_{av}$, deg/in.-lb	0.036

9.8   Relationship Between Sideslip Angle and Fin Angle of Attack

The relationship between sideslip angle and fin angle of attack must be considered. Vertical tail loads may be defined as a function of fin angle of attack or airplane sideslip angle, as shown in Eqs. (9.21) and (9.22):

$$L_v = C_{Lv\alpha} q S_v \alpha_v \qquad (9.21)$$

$$L_v = C_{Lv\beta} q S_v \beta \qquad (9.22)$$

The relationship between fin angle of attack and airplane sideslip angle[2] is

$$\alpha_v = \sigma - \beta = \left(1 - \frac{d\sigma}{d\beta}\right)(-\beta) \qquad (9.23)$$

where α_v is the vertical tail angle of attack (deg), β is the airplane sideslip angle (deg), σ is the angle of sidewash analogous to the downwash angle ϵ (deg), $C_{Lv\alpha}$ is the change in vertical tail lift coefficient per change in fin angle of attack (per deg), and $C_{Lv\beta}$ is the change in vertical tail lift coefficient per change in airplane sideslip angle (per deg).

Substituting Eq. (9.23) into Eq. (9.21) and equating this to Eq. (9.22), one can obtain the following relationship:

$$C_{Lv\beta} = -\left(1 - \frac{d\sigma}{d\beta}\right)C_{Lv\alpha} \qquad (9.24)$$

The sidewash parameter $(1 - d\sigma/d\beta)$ at the vertical tail, shown in Fig. 9.8 as a function of wing angle of attack and Mach number, was obtained from wind-tunnel tests.

9.8.1   Effect of Sidewash on Vertical Tail Loads

The usual practice for load analyses is to obtain the spanwise pressure distribution from the wind tunnel due to sideslip angle β. The data then are integrated to obtain the vertical tail coefficient $C_{Lv\beta}$ at the various Mach numbers tested.

Because the fin is a fixed surface, the vertical tail lift variation with fin angle of attack requires a special test setup in the wind tunnel. The vertical tail is rotated on the body at various fin angles of attack, similar to the testing for the horizontal tail, to obtain the load variation with stabilizer angle of attack.

For rudder and engine-out maneuvers, the load relationships with respect to sideslip are used as the maneuvers are defined in terms of sideslip and rudder angles.

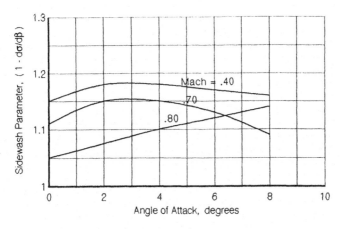

Fig. 9.8   Sidewash parameters at the vertical tail.

For lateral gust load analyses, defined in terms of the variation in fin angle of attack due to gust penetration, the resulting vertical tail loads are usually based on lift coefficients per sideslip angle as shown in Eq. (9.25):

$$Lv_{\text{gust}} = C_{Lv\beta} q S_v f(\alpha_v)_{\text{gust}} \tag{9.25}$$

Substituting Eq. (9.24) into Eq. (9.25) will result in the following relationship, neglecting the minus sign ahead of the sidewash parameter:

$$Lv_{\text{gust}} = \left(1 - \frac{d\sigma}{d\beta}\right) C_{Lv\alpha} q S_v f(\alpha_v)_{\text{gust}} \tag{9.26}$$

Thus the resulting gust loads are conservative by the factor $(1 - \sigma_\beta)$ since this parameter is usually greater than 1, as shown in Fig. 9.8.

This conservatism is usually included in the lateral gust dynamic analyses because the vertical tail loads database used for rudder maneuver analyses is not corrected for the sidewash parameter.

References

[1]Anon., "Part 4b—Airplane Airworthiness Transport Categories," Civil Air Regulations, Civil Aeronautics Board, Dec. 1953.

[2]Etkin, B., *Dynamics of Flight, Stability and Control,* Wiley, New York, 1959.

10
Wing Loads

The determination of wing loads for structural design is important, of course, not only from a structural adequacy point of view but also from the impact on structural weight. Wing aeroelasticity significantly affects wing loads and horizontal tail balance loads, thus impacting the structural weight of the aft body, horizontal tail, and wing structure.

With the advent of swept wings, aeroelasticity has become a major factor in determining the spanwise distribution of loads, both from a static and dynamic point of view. Structural deflections of the wing tips under limit design loads vary from 5 ft for the 737-200 models to almost 22 ft for the 747-400 models.

10.1 Wing Design Criteria

The design criteria for the basic wing box structure are based on the criteria previously discussed in Chapters 2–7. An additional criterion that will be discussed is the application of unsymmetrical gusts as specified in FAR 25.349(b).

10.2 Wing Design Conditions

Consideration will be given to the conditions that contribute to the structural design of the wing structural box before discussing the methods of analysis for calculating wing design loads. For the purpose of this discussion the design loads on the wing along a specified load reference axis will be assumed to be represented by net beam shear, beam-bending moment, and torsion as shown in Fig. 10.1. Chordwise shear and bending moment as shown in this figure will be discussed in a later section.

10.2.1 Wing Static Load Envelopes

Wing load envelopes in terms of design shear, bending moment, and torsion are shown in Figs. 10.2–10.4 for flight, landing, and ground-handling static load conditions for a typical commercial jet transport. These conditions are discussed in Secs. 10.3–10.5.

Dynamic gust, landing, and taxi conditions are discussed in Secs. 10.7–10.9.

10.3 Symmetrical Maneuver Analysis

The symmetrical maneuver requirements and analyses discussed in Chapter 2 are solved to determine the required parameters necessary for calculation of wing loads.

These parameters are 1) airplane load factor n_z specified in Tables 2.1 and 2.2, 2) pitching acceleration about the airplane center of gravity $\ddot{\theta}$ (rad/s^2), 3) pitching velocity $\dot{\theta}$ (rad/s), 4) wing reference angle of attack α_w (deg), 5) inertia parameters due to operating empty weight (OEW) and fuel, and 6) airspeed and Mach number.

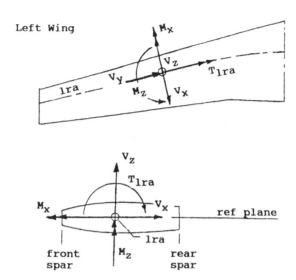

Fig. 10.1   Nomenclature and sign convention for wing structural box loads: *lra* = load reference axis; M_x = beam bending moment, in. lb; M_z = chordwise bending moment, in. lb; T_{lra} = torsion about the load reference axis, in. lb; V_x = chordwise shear, lb; V_y = axial shear, lb; V_z = shear normal to reference surface, lb.

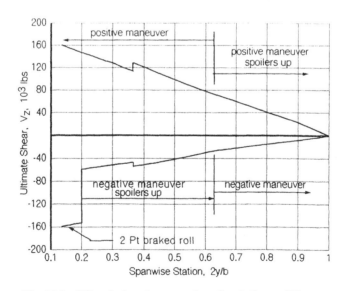

Fig. 10.2   Wing design shear envelope for static conditions.

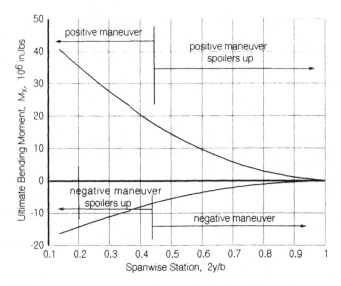

Fig. 10.3   Wing design bending moment envelope for static conditions.

The wing spanwise load distribution may be considered the sum of the increments as shown in Eq. (10.1):

$$\{l\} = \{l_\alpha\}\alpha_w + \{l_0\} + \{l_n\}n_z + \{l_{\dot\theta}\}\dot\theta + \{l_{\ddot\theta}\}\ddot\theta + \{l_{\rm sp}\} \tag{10.1}$$

where $\{l\}$ is the net spanwise lift distribution (lb/in.), $\{l_\alpha\}\alpha_w$ is the lift distribution due to angle of attack (lb/in.), $\{l_0\}$ is the lift distribution at $\alpha_w = 0$ (lb/in.), $\{l_n\}n_z$ is the lift distribution due to aeroelastic effect of inertia, equal to zero for a rigid wing (lb/in.), $\{l_{\dot\theta}\}\dot\theta$ is the lift distribution due to pitching velocity (lb/in.), $\{l_{\ddot\theta}\}\ddot\theta$ is the lift distribution due to aeroelastic effect of inertia, equal to zero for a rigid

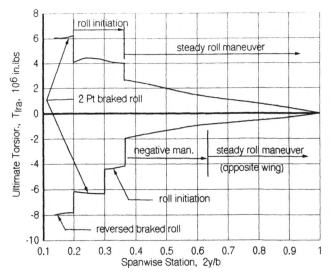

Fig. 10.4   Wing design torsion envelope for static conditions.

wing (lb/in.), and $\{l_{sp}\}$ is the lift distribution due to speedbrakes (spoilers or other symmetrically deflected devices).

Equation (10.1) applies to either a flaps-up or flaps-down analysis, whereby flap deflection is included in the first two terms. The effects of aeroelasticity are included in all terms if a flexible analysis is used.

10.3.1 Steady-State Maneuvers

In general, the steady-state maneuver conditions shown in Fig. 2.1 will produce the maximum design wing loads for symmetrical maneuvers. For these conditions pitching acceleration is zero.

Wing shear, bending moment, and torsions are shown in Figs. 10.2–10.4 for the symmetrical flight maneuver conditions that are part of the design load envelope for a typical wing. The critical maneuver conditions are shown for positive and negative load factors with spoilers acting as speedbrakes, extended and retracted. The effect of speedbrakes extended is discussed in Sec. 10.10.

Flaps-down maneuver conditions in general are not critical for wing-bending moment, but the rear spar is critical for the trailing-edge flap support loads and the associated shear and torsion.

The steady-state maneuver wing angle of attack may be calculated using Eq. (2.21) or by direct solution of the symmetrical maneuver analyses discussed in Sec. 10.14.2.

For symmetrical maneuver conditions where pitching acceleration is zero, pitching velocity may be calculated for curvilinear flight using Eq. (10.2):

$$\dot{\theta} = (n_z - 1)g/V \qquad \text{(rad/s)} \qquad (10.2)$$

where n_z is the steady-state maneuver load factor, $g = 32.2$ ft/s², and V is the airplane true airspeed (ft/s).

The wing angle of attack due to pitching velocity is calculated from Eq. (10.3):

$$\{\blacktriangle\alpha_w\}_{\dot{\theta}} = [(n_z - 1)g/V]\{x_V - x_{cg} + c/2\} \qquad \text{(rad)} \qquad (10.3)$$

where $\{\blacktriangle\alpha_w\}_{\dot{\theta}}$ is the spanwise angle of attack due to pitching velocity (rad).

The geometric parameters in Eq. (10.3) are defined in Fig. 10.5 for a typical aerodynamic analysis station.

10.3.2 Abrupt Unchecked Pitch Maneuver

For aircraft certified before 1978, the abrupt unchecked pitch maneuver was not considered a critical condition for wing loads because the maneuver was essentially a level flight condition at 1-g load factor.

With the change in FAR 25.331(c)(1) by Amendment 25-46 in December 1978, the response of the airplane must be considered when calculating tail loads. Airplane loads that occur after reaching the design maneuver load factor need not be considered.

The question can be asked, what happens to wing loads when the design load factor is reached during the abrupt unchecked pitch condition? Because for a given altitude the unchecked pitch maneuver is accomplished at V_A speed, [see FAR 25.331(c)(1)], the maximum load factor obtainable would be less than the design maneuver load factor for aircraft where compressibility effects are considered in determining $C_{N max}$. An analysis is shown in Table 10.1 where the resulting load

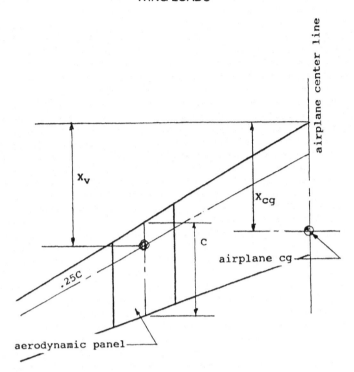

Fig. 10.5   Wing section angle of attack due to pitch velocity.

factor is less at V_A speeds than for the steady-state maneuver at the design load factor. This condition is not considered applicable as a wing design condition.

10.3.3 Abrupt Checked Maneuvers

Abrupt checked maneuvers are usually not considered as critical wing conditions. The basic reason is the reduction in wing angle of attack at the time that maximum maneuver load factor is attained during the checked maneuver.

Table 10.1   Maneuvering capability at V_A speeds compared
with the airspeed for the upper left-hand corner
of the design V-n diagram[a]

	n_z	C_{Nmax}	q, lb/ft^2	V_e, keas	Mach
1-g stall condition	1.0	1.163	112.12	181.9	0.275
V_A speed[b]	2.36	1.096	280.79	287.8	0.435
+HAA speed[c]	2.50	1.096	297.45	296.2	0.448

[a]Airplane parameters are assumed to be $S_w = 2500$ ft^2, Altitude = sea level, and $W = 326,000$ lb.
[b]$V_A = V_{s1g}\sqrt{n_z}$ [Eq. (15.6)], and V_{s1g} is the stall speed at $n_z = 1.0$ (keas).
[c]+HAA speed = airspeed at upper left-hand corner of the V-n diagram (see Fig. 14.7).

Consider the net horizontal tail load during the checked maneuver as made up of two parts:

$$L_{t\text{cm}} = BTL + \blacktriangle L_{t\text{cm}} \tag{10.4}$$

where BTL is the balancing tail load increment (lb), and $\blacktriangle L_{t\text{cm}}$ is the incremental tail load due to checked maneuver (lb).

For a symmetrical steady-state maneuver, the lift balance equation is shown in Eq. (10.5) relating the tail-off lift L_{to} and balancing tail load to the inertia term $n_z W$:

$$n_z W = L_{\text{to}} + BTL \tag{10.5}$$

Balancing tail load BTL may be defined as shown in Eq. (10.6) using the linear analysis represented by Eqs. (2.22) and (2.24):

$$BTL = BTL_0 + \left(\frac{dBTL}{dn_z} \right) n_z \tag{10.6}$$

Combining Eqs. (10.4)–(10.6), one can derive the lift balance equation for the checked maneuver:

$$n_z W = L_{\text{to}} + BTL_0 + \left(\frac{dBTL}{dn_z} \right) n_z + \blacktriangle L_{t\text{cm}} \tag{10.7}$$

If the load factor at the time of peak checked maneuver tail load is to not exceed the design load factor, then using Eq. (10.7), one must reduce the tail-off lift to compensate for the added incremental tail load $\blacktriangle L_{t\text{cm}}$. Hence,

$$L_{\text{to cm}} < L_{\text{to bal man}} \tag{10.8}$$

where $L_{\text{to cm}}$ is the tail-off lift at time of peak checked maneuver tail load, and $L_{\text{to bal man}}$ is the tail-off lift for the steady-state balanced maneuver.

Based on this rationale, the wing loads for the checked maneuver conditions should be less than the wing loads obtained during the steady-state maneuver.

The pitching acceleration at the time of the checked maneuver will be negative for a positive load factor and will contribute an incremental positive load factor acting on the wing inertia for that portion of the wing aft of the airplane center of gravity position. For the inboard wing that may be forward of the airplane center of gravity, the opposite is true. The effect of pitching acceleration on the resulting wing maneuver loads increases the relief due to inertia loads. The aeroelastic effect on the resulting wing angle of attack distribution must be given consideration.

10.4 Rolling Maneuver Analysis

The rolling maneuver analysis requirements and equations of motion presented in Chapter 3 are solved to determine the required parameters necessary for calculation of wing loads. These parameters are 1) airplane load factor n_z specified in Table 3.2 and the resulting symmetrical flight wing loads, 2) maximum roll velocity $\dot{\phi}$ for the steady roll condition, and 3) maximum roll acceleration $\ddot{\phi}$ and related roll velocity for the roll acceleration condition.

Roll parameters can be determined using the simplified analysis calculated using Eqs. (3.5) and (3.6) or from the solution of the equation of motion discussed in Sec. 3.4.

At the time of publication of this book, the commercial regulations did not require a specific recovery condition; therefore it has been assumed that the recovery is made such that the resulting wing loads are less critical than the roll initiation condition.

The critical wing roll maneuver loads for the examples shown in Figs. 10.2–10.4 are critical for the wing box structure in the vicinity of the ailerons that are located outboard of 75% of the wing span. However, rolling maneuvers do make up part of the torsion envelope for the inboard wing as shown in Fig. 10.4.

The airplane configuration for the loads in Figs. 10.2–10.4 has ailerons and spoilers for roll control. The symmetrical flight loads are rebalanced to maintain a maneuver load factor of 1.67 during the roll as discussed in Sec. 10.4.2.

10.4.1 Ailerons and Spoilers Used for Lateral Control

The lateral control available as a function of pilot control wheel angle is shown in Fig. 3.5 for a typical control configuration. Spoilers usually have a delay in the system to minimize autopilot trim characteristics during normal cruise flight. In this example spoilers are not used until the control wheel angle is greater than 10 deg.

10.4.2 Symmetrical Load Increments

Wing loads for the symmetrical maneuver increments are calculated for airplane load factors of zero and two-thirds of the positive maneuvering load factor used for design.

For aircraft configurations with unsymmetrical operation of lateral control devices, a correction must be made to maintain the design load factor during the roll.

10.4.2.1 Unsymmetrical operation of ailerons.   Because of the design characteristics of the lateral control system, one wing may have more up than down aileron available, particularly at extreme wheel positions as shown in Fig. 10.6.

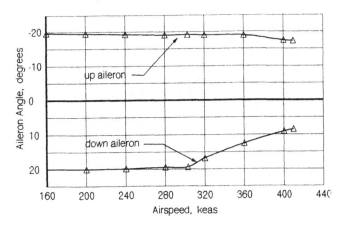

Fig. 10.6   Maximum aileron available under load.

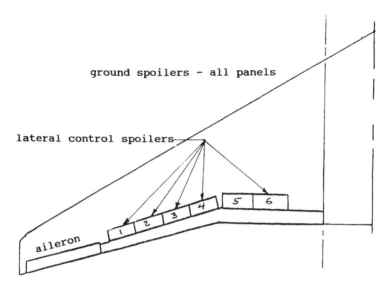

Fig. 10.7   Spoilers used for lateral control.

This will produce an unbalanced lift distribution across the wing that will require an adjustment to the symmetrical maneuver wing angle of attack to maintain the desired maneuver load factor during the roll.

10.4.2.2   Use of spoilers for lateral control.   Wing spoilers, used for lateral control as shown in Fig. 10.7, present a particular problem in solving for the resulting wing loads during a rolling maneuver. Roll conditions must be considered in both the clean wing configuration, whereby spoilers are not extended as speedbrakes, and for speedbrakes-extended conditions. The spoilers are operated differentially such that the contribution to roll may be from reduction of spoilers on one wing vs extension on the other wing.

   In either case a loss of lift will occur that must be corrected by increasing the wing angle of attack during the maneuver to maintain the required symmetrical load factor.

10.4.3   Spanwise Load Distributions During Rolling Maneuvers

   The spanwise incremental load distributions during rolling maneuvers are depicted in Fig. 10.8 for the left and right wings. Lateral control devices and roll damping include the contribution of aeroelasticity if a flexible wing analysis is used. The lift distribution due to roll acceleration is due to aeroelasticity.

   The unsymmetrical loads acting on the wings during rolling maneuvers are the net sum of the increments shown depending on the roll condition under investigation.

10.4.4   Rolling Maneuver Load Factors

   The load factors acting on the wing during rolling maneuvers are shown in Figs. 10.9 and 10.10 for the maximum roll acceleration and steady roll conditions for an assumed aircraft. The roll acceleration load factors as shown are assumed normal to the wing and are additive to the symmetrical maneuver factor on one wing and subtractive on the opposite wing. Load factors for the steady roll condition acting

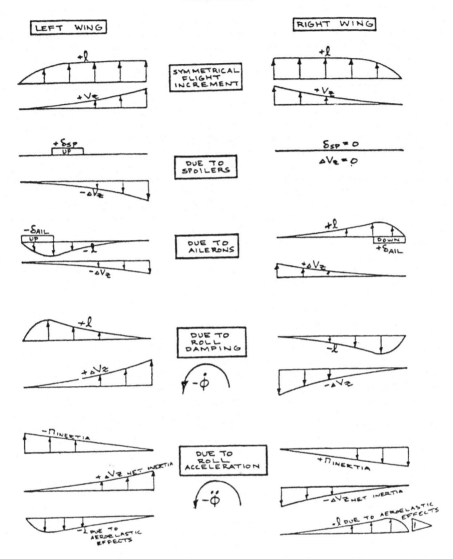

Fig. 10.8 Incremental airload for rolling maneuvers.

outboard on both wings must be combined with the symmetrical maneuver load factors.

The following equations summarize the load factors during rolling maneuvers acting on the wing structure and contents, including external stores such as wing-mounted engines.

Left wing:

$$n_z = n_{z\,\text{sym}} - y\ddot{\phi}/g \qquad\qquad (10.9)$$

Right wing:

$$n_z = n_{z\,\text{sym}} + y\ddot{\phi}/g \qquad\qquad (10.10)$$

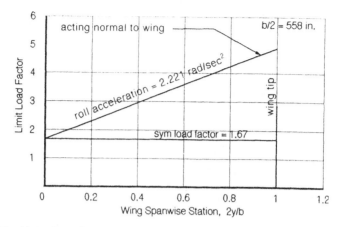

Fig. 10.9   Load factor spanwise distribution due to roll acceleration.

Both wings (acting outboard):

$$n_y = y\dot{\phi}^2/g \qquad (10.11)$$

where y is the spanwise distance from the centerline of the airplane (in.), and g is the acceleration of gravity (in./s²).

Wing-mounted nacelle load factors due to rolling maneuvers as obtained from a time history analysis are compared in Fig. 10.11 with the envelope of conditions defined using the simplified analysis defined by Eqs. (3.5) and (3.6).

10.4.5   Flight Testing of Rolling Maneuver Conditions

The effect of the loss of lift due to spoilers extended for lateral control during rolling maneuvers has always been difficult to overcome during flight load survey tests. The pilots are able to set up the initial conditions from wind-up turns that would be a steady banked turn of 53.2 deg for a load factor of 1.67 g, as calculated from Eq. (3.2). As the rolling maneuver progresses, the airplane load factor reduces primarily due to the loss of lift caused by spoilers, thus requiring the pilot to pull

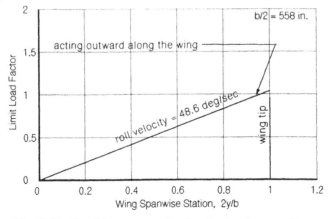

Fig. 10.10   Load factor spanwise distribution due to roll velocity.

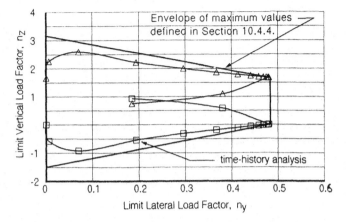

Fig. 10.11   Nacelle load factors due to rolling maneuver.

the airplane nose up to maintain constant load factor during the maneuver. Usually the condition requires much practice before the maneuver can be accomplished.

The question arises whether the maneuver is a valid design condition when the pilot would not compensate for the loss of lift due to unsymmetrical lateral control devices. Again this is still a criterion, and although it requires some correction in the symmetrical analysis, the resulting load condition may not be significantly more critical.

Flight testing of zero load factor conditions for rolling maneuvers is very difficult and is usually not attempted during the flight load survey programs.

10.5 Yawing Conditions

The yawing maneuver requirements discussed in Chapter 4 and the lateral gust requirements discussed in Chapter 5 involve design conditions that are critical for empennage and fuselage structure. In general, the wing structure is not critical for these types of conditions, except for the attachment of engine/nacelles located outboard on the wing or other such external store devices located on the wing. Winglets are critical for these maneuvers and must be given special consideration.

The need for compatible load conditions on the wing for yawing maneuver loads on nacelles (or gust loads) can be of importance when modeling the nacelle and local wing structure by a finite element analysis.

10.5.1 Wing Loads in Yawed Flight

The aerodynamic moments on the airplane about the airplane roll axis may be written in coefficient form as shown in Eq. (10.12):

$$C_l = C_{l\beta}\beta + C_{l\dot\phi}\dot\phi + C_{l\dot\psi}\dot\psi + C_{lsp} + C_{lail} + C_{lrud} \qquad (10.12)$$

Yawing maneuvers are assumed to be accomplished with the wings held in a level flight attitude; hence the roll rate $\dot\phi$ is assumed to be zero. If the sideslip angle is maximum for the condition being investigated, then the yaw rate $\dot\psi$ will also be zero at that time.

Equation (10.12) may be rewritten as shown in Eq. (10.13) where the net rolling moment coefficient is zero, thus complying with the assumption of a flat maneuver:

$$C_{l\beta}\beta + C_{lsp} + C_{lail} + C_{lrud} = 0 \qquad (10.13)$$

Therefore the loads acting on the wing in a yawing maneuver at the time of maximum sideslip angle for the condition under investigation are a function of only the sideslip angle and the lateral control applied to maintain the wings-level attitude. The level flight ($n_z = 1.0$) symmetrical flight wing loads must be added to the incremental loads due to the yawing maneuver.

If the incremental spanwise pressure distributions on the left- and right-hand wings due to sideslip have been obtained from wind-tunnel tests, then the load distributions over the wing in yawed flight may be calculated directly from integrated data.

The incremental spanwise pressure distribution due to sideslip as discussed in the previous paragraph would be for a rigid wing. Aeroelastic corrections could be made by using the ratio of the elastic wing/rigid wing spanwise symmetrical load distribution due to wing angle of attack.

10.5.2 Approximation of Wing Loads in Yawed Flight

An approximation of the incremental wing loads due to sideslip may be obtained in the following manner, if spanwise pressure data are not available from wind-tunnel or flight test measurements.

1) Obtain rolling moment due to sideslip for the airplane from tail-off wind-tunnel data.

2) Assume no contribution of rolling moment due to body or external stores on the wing:

$$(C_{l\beta}\beta)_{\text{wing}} = (C_{l\beta}\beta)_{\text{tail-off}} \qquad (10.14)$$

3) Assume that the distribution of loads on the left and right wings are represented by the sweep parameters shown in Fig. 10.12, where

$$\Lambda_{\beta lw} = \Lambda_{0.25} + \beta \qquad (10.15)$$
$$\Lambda_{\beta rw} = \Lambda_{0.25} - \beta \qquad (10.16)$$

The ratio of the values obtained from Fig. 10.12 is used to calculate the load distribution on the left and right wing due to sideslip from the equivalent symmetrical airload distribution.

By determining the value of the term $((\beta C_{L\alpha}/k)$ for the two effective sweep angles using Eqs. (10.15) and (10.16) and the value for the baseline sweep, $\Lambda_{0.25}$, one can calculate the ratio for the left and right wings:

$$R_{lw} = (\beta C_{L\alpha}/k)_{lw}/(\beta C_{L\alpha}/k)_{\text{baseline}} \qquad (10.17)$$
$$R_{rw} = (\beta C_{L\alpha}/k)_{rw}/(\beta C_{L\alpha}/k)_{\text{baseline}} \qquad (10.18)$$

The incremental airload in yawed flight may be approximated as shown in Eqs. (10.19) and (10.20), using the parameters calculated from Eqs. (10.17) and (10.18):

$$\{L_\beta\}_{lw} = R_{lw}(L_\alpha)_{\text{sym}}R_{\text{cor}} \qquad (10.19)$$

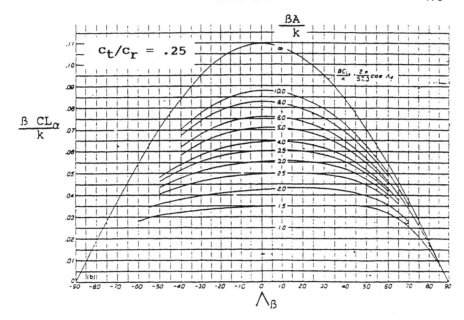

Fig. 10.12   Lift curve slope variation with wing sweep parameters for symmetrical flight [NACA Report 921 (Ref. 1)]: A = aspect ratio = b^2/S_w; $\beta = \sqrt{1 - M^2}$; $\Lambda_\beta = \tan^{-1}$ $(\tan \Lambda/\beta)$; k = ratio of experimental section lift curve slope to $(2\pi/\beta)$.

$$\{L_\beta\}_{rw} = R_{rw}(L_\alpha)_{sym} R_{cor} \tag{10.20}$$

where $(L_\alpha)_{sym}$ is the symmetrical flight shear, moment, or torsion along the load reference axis due to airload,

$$R_{cor} = (C_{l\beta})_{wind\ tunnel}/(C_{l\beta})_{uncorrected} \tag{10.21}$$

$$(C_{l\beta})_{uncorrected} = 2C_{L\alpha}y_{cp}(R_{lw} - R_{rw})/b \qquad (per\ deg) \tag{10.22}$$

where $C_{L\alpha}$ is the lift curve slope of the wing (per deg), y_{cp} is the spanwise center of pressure for symmetrical airload (in.), b is the wing span (in.), lw is the left wing, and rw is the right wing.

The lift curve slope variation with wing sweep parameters for symmetrical flight conditions as shown in Fig. 10.12 is obtained from Ref. 1.

The net loads in yawed flight using the assumptions discussed in Sec. 10.5.1 are calculated from Eqs. (10.23) and (10.24):

$$\{L\}_{net\ lw} = \{L\}_{sym} + \{L_\beta\}_{lw} + \{L\}_{ail} \tag{10.23}$$

$$\{L\}_{net\ rw} = \{L\}_{sym} + \{L_\beta\}_{rw} + \{L_{ail}\}_{rw} + \{L_{sp}\}_{rw} \tag{10.24}$$

where $\{L_{...}\}$ is the incremental shear, moment, or torsion along load reference axis; sym is the symmetrical 1-g flight net loads; sp is the incremental airload due to spoilers; ail is the incremental airload due to ailerons; and β is the incremental airload due to sideslip.

10.5.3 Yawing Maneuver and Lateral Gust Conditions

Yawing maneuver and lateral gust conditions are usually not critical for the wing structure outboard of the side of the body except for the design of the engine/nacelle support structure for engines mounted on the wings. This also applies to other external stores such as wing tank pods, engine pods mounted on the wing for ferry purposes, and winglets mounted on the wing tips.

The wing center section may be critical for yawing maneuver and lateral gust conditions.

10.6 Landing and Ground-Handling Static Load Conditions

The wing conditions for the landing and ground-handling static load requirements discussed in Chapters 6 and 7 are considered in this section. The two conditions that make up part of the static load envelopes shown in Figs. 10.2 and 10.4 are the two-point braked-roll condition discussed in Sec. 7.4.2 and the reversed braking condition discussed in Sec. 7.4.5. These conditions are critical in shear and torsion for the wing box inboard of the main landing gear that is mounted on the wing.

The two-point landing conditions, level landing, and tail-down landing defined in Sec. 6.3 are critical for the main gear and related support structure on the wing. Both the spin-up and spring-back conditions are considered in the design of this structure.

Dynamic landing and taxi conditions are discussed in Secs. 10.8 and 10.9, respectively.

10.6.1 Wing Loads Compatible with Main Gear Ground-Handling Design Conditions

Wing loads for the main landing gear ground-handling conditions discussed in Chapter 7 are readily computed using the applicable load factors specified for

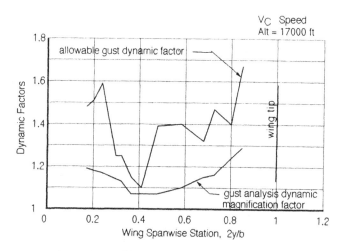

Fig. 10.13   Wing margins for gust dynamics.

each condition. These load factors are applied to the inertia loads for the condition under investigation with the appropriate maximum or minimum fuel load at the gross weight under consideration.

Wing airload is assumed zero for these conditions.

10.6.2 Wing Loads Compatible with Main Gear Landing Conditions

Wing loads for the main gear design landing conditions must include airload and inertia loads.

Load factors compatible with the main gear design level and tail-down landing loads may be calculated as discussed in Sec. 6.3. The airplane is placed in balance using the appropriate pitching accelerations for the spin-up and spring-back conditions.

The inertia loads due to landing impact are added to the 1-g flight condition.

10.7 Gust Loads and Considerations for Dynamics

The application of the vertical gust requirements discussed in Chapter 5 to wings on commercial aircraft certified in 1958 to the present generation of airplanes has evolved over the intervening years due to criteria changes and the introduction of modern computer technology.

10.7.1 Historical Perspective on the Consideration for Dynamics

Because of the lack of sophisticated computers before 1960 that were capable of handling multiple degrees of freedom, dynamic gust analyses were assessed for only a few flight conditions. Maximum loads such as bending moments, shear, and torsion were obtained from simple digital computer analyses or solved directly on an analog computer. Time histories were obtained for the one-minus-cosine gust shape, and at times some tuning was accomplished.

One of the simplest methods used in assessing the effect of dynamics on wing structural loads was to calculate an allowable gust dynamic factor using Eq. (10.25):

$$DF = (Mx \text{ design} - Mx \, 1g)/(Mx \text{ static gust} - Mx \, 1g) \qquad (10.25)$$

where Mx design is the wing design limit bending moment (in.-lb), $Mx \, 1g$ is the wing 1-g flight limit bending moment (in.-lb), and Mx static gust is the wing static gust limit bending moment (in.-lb).

Wing static gust loads were calculated by the gust formula method using Eqs. (5.4) and (5.14).

Dynamic magnification factors (DMF) were computed from the available dynamic simulations using bending moments as shown in Eq. (10.26). Static elastic bending moments were calculated from the same database as that used in the dynamic analysis. Both analyses were calculated using a discrete 25-chord one-minus-cosine gust shape:

$$DMF = (\Delta Mx_{\text{dynamic}})/(\Delta Mx_{\text{static elastic}}) \qquad (10.26)$$

A comparison of the allowable gust dynamic factor determined from Eq. (10.25) with the dynamic magnification factor calculated from Eq. (10.26) at various spanwise stations is shown in Fig. 10.13. Before 1960 this approach was deemed adequate for certification by the U.S. Federal Aviation Administration (FAA) and British Civil Aviation Authority (CAA).

During the period from 1960 to 1965 the effect of continuous turbulence on aircraft structure was considered. These loads were usually compared with the design maneuver load envelopes and were not actually used in any stress analysis of the wing box structure.

10.7.2   Application of Full Dynamic Analysis for Gust Loads

After 1965, the certifying agencies required a full dynamic structural assessment of the effect of gust loads on the aircraft. With the availability of more sophisticated digital computers capable of handling multiple degree-of-freedom analyses, both discrete and continuous turbulence solutions were accomplished.

Solutions for vertical gust loads due to discrete and continuous turbulence are presented in Refs. 6, 7, and 9 of Chapter 5. The equations of motion are developed and methods of analysis for solving the discrete and continuous turbulence analyses are discussed in detail in these references.

Three types of analyses are discussed as representative of the wing loads resulting from dynamic gust modeling. These examples are for a wing that is generally maneuver critical, particularly in terms of wing-bending moment. The combination of shear and torsion shown in these figures may contribute to critical spar shear flows.

10.7.2.1   Discrete gust analysis per FAR/JAR 25.341(a).   Wing loads due to the discrete gust analysis criteria of the FAR/JAR 25.341(a) harmonization process are compared in Fig. 10.14 with the design flight load envelopes for a narrow-body freighter aircraft. The design gust velocities are determined from Fig. 5.2 as modified by the flight profile alleviation factors shown in Table 5.4.

The structural dynamic response is included in the calculation of the incremental gust the loads represented by shear, moment, and torsion. The resulting incremental gust loads are combined with 1-g flight loads with and without speedbrakes extended as discussed in Sec. 10.10.

The critical gust gradients for wing-bending moment, torsion, and front spar shear flow are compared in Table 10.2 for the discrete gust condition shown in Fig. 10.14. This comparison shows that the gradients for shear flow differ significantly from the values shown for torsion. In general, torsion maximums are

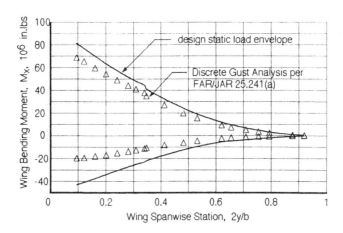

Fig. 10.14   Wing loads for discrete vertical gust dynamic analysis.

Table 10.2   Comparison of gust gradients for maximum bending, torsion, and front spar shear flow, discrete vertical gust analysis[a]

	Gust gradient at maximum load shown, ft		
$2y/b$	Positive bending moment	Positive torsion	Front spar shear flow
0.876	83	333	83
0.793	333	62	125
0.532	250	42	125
0.412	250	42	125
0.308	250	125	333
0.240	208	125	333
0.160	208	125	83

[a]Shear flows due to positive shear and torsion are additive for the front spar.

not a good indication of criticality, but rather the selection of critical gradients should also be made considering front and rear spar shear flows.

10.7.2.2   Continuous turbulence design envelope analysis.   Wing loads for the continuous turbulence requirements of FAR 25, Appendix G(b), are shown in Fig. 10.15 for the design envelope analysis using the minimum design gust velocities. The minimum gust values, shown in Fig. 5.4, have been accepted for the illustrative airplane because of the similarity with existing designs with extensive satisfactory service experience.

The structural dynamic response is included in the calculation of the values of $\bar{A}$ for the loads represented by shear, moment, and torsion. The power spectral density of the atmospheric turbulence is represented by Eq. (5.10) with $L = 2500$ ft. The resulting incremental gust loads are combined with 1-g flight loads with and without speedbrakes extended as discussed in Sec. 10.10.

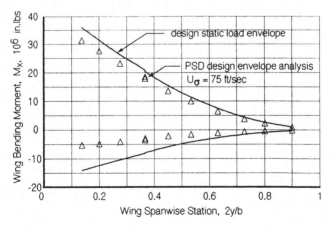

Fig. 10.15   Wing loads for continuous turbulence vertical gust dynamic analysis design envelope conditions.

Continuous turbulence analysis net loads are compared in Fig. 10.15 with the static design load envelopes discussed in Sec. 10.2.1. These gust loads are less critical than the maneuver loads for the airplane configuration shown in this example.

10.7.2.3   Continuous turbulence mission analysis.   Wing loads for the continuous turbulence requirements of FAR 25, Appendix G(c), mission analysis, are also shown in Fig. 10.16.

The missions include climb, cruise, and descent segments as necessary to represent three typical flight lengths of the aircraft in service.

The structural dynamic response is included in the calculation of the values of $\bar{A}$ for the loads represented by shear, moment, and torsion. The power spectral density of the atmospheric turbulence is represented by Eq. (5.10) with $L = 2500$ ft. Limit gust loads were determined at a frequency of exceedance of 2×10^{-5} exceedances per hour. Both positive and negative gust loads are considered in determining limit loads.

10.8   Wing Loads for Dynamic Landing Analysis

Wing loads are compared for a dynamic landing analysis in Fig. 10.17 with the static design load envelope. All appropriate structural modes are included in this analysis.

Lift equal to the airplane gross weight is assumed acting on the airplane throughout the oleo stroke during the landing contact with the runway. Appropriate airspeeds for the level landing with the nose gear just off of the ground or the taildown condition are determined as discussed in Sec. 6.2.

As noted in Fig. 10.17, the shear and torsion inboard of the nacelle exceed the static load envelope. For this condition time-phased loads are provided for stress evaluation of the dynamic landing analysis conditions.

10.9   Wing Loads for Dynamic Taxi Analysis

Wing loads are calculated for a dynamic taxi analysis using the San Francisco runway no. 28R as defined in Fig. 7.3. This runway roughness description, before

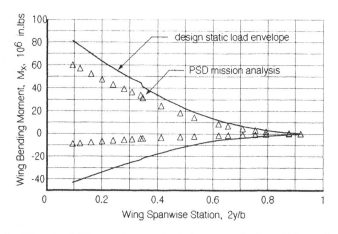

Fig. 10.16   Wing loads for continuous turbulence vertical gust dynamic mission analysis conditions.

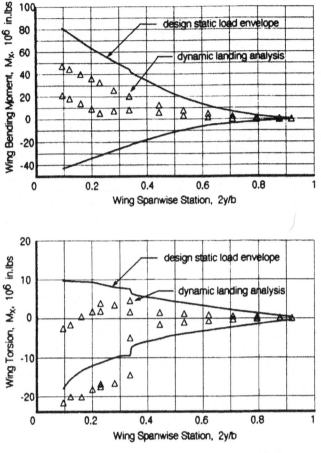

Fig. 10.17   Wing loads for dynamic landing analysis.

refurbishment, is considered acceptable for meeting the requirements of FAR/JAR 25.491.

The airplane is taxied over the runway at various speeds, and then an analytical takeoff is performed (in both directions) to obtain the maximum loads applied to the aircraft structure. Airloads are applied to the flight structure during the analytical takeoffs at the appropriate takeoff flap settings. Structural modes representing the wing, body, nacelles, and main landing gears are included in the analysis. The shock-absorbing mechanism of the landing gear oleo system is represented in the analysis.

The resulting wing loads are shown in Fig. 10.18 for a typical airplane dynamic taxi analysis. As noted, the loads are significantly lower for this example than for the wing static design load envelope discussed in Sec. 10.2.1.

10.10   Effect of Speedbrakes on Symmetrical Flight Conditions

The effect of speed control devices, such as wing spoilers, must be considered for symmetrical maneuvers per the requirements of FAR 25.373. If wing-mounted spoilers are used as speedbrakes, the wing spanwise lift distribution will be modified as shown in Fig. 10.19.

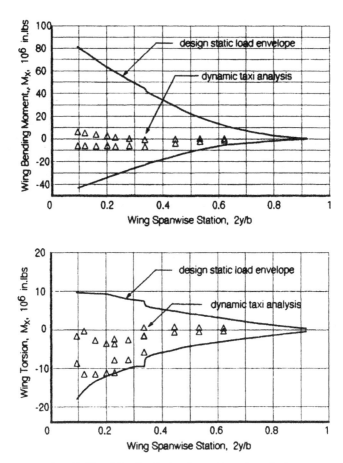

Fig. 10.18   Wing loads for dynamic taxi analysis.

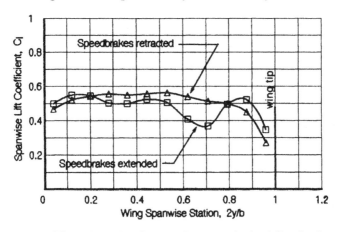

Fig. 10.19   Effect of speedbrakes on wing spanwise load distribution.

To compensate for the loss in lift due to spoilers during positive maneuver conditions, the wing angle of attack must be increased to maintain flight at a given load factor. This has the effect of increasing the wing shear and bending moment outboard of the spoilers and hence will be more critical than the speedbrakes-retracted conditions. This is shown in Figs. 10.2 and 10.3.

For negative maneuver conditions, the opposite will happen. Wing loads will be more critical inboard for spoiler-up conditions and outboard for spoiler-retracted maneuvers.

Speedbrakes extended must be included in the symmetrical flight 1-g load conditions for the vertical gust analysis loads in a manner similar to that of the design maneuver conditions.

10.11 Effect of Fuel Usage on Wing Loads

The effect of fuel usage must be considered in determining the spanwise distribution of net wing loads that are the sum of airloads and inertia loads. If the airplane has multiple tanks in both the wing and body, then fuel usage may have a profound effect on the resulting design loads.

In the design of a modern narrow-body airplane, the placement of the wing fuel tanks was studied to optimize the load relief due to inertia such that the fuel was consumed from the center wing tanks before the outboard wing fuel was used. This required fuel pumps to be placed in the center tank to continuously maintain full fuel in the outboard tanks until the center tanks emptied. The inboard tank end rib position was selected on the basis of this optimization study, as shown in Fig. 10.20. shown in Fig. 10.21 for an airplane with a similar fuel tank arrangement. If the outboard tanks had been larger and the center tank smaller, as was originally proposed, the outboard wing-bending moments would be higher than the final design moments.

If the wing center tank fuel is retained while significant outboard wing fuel is used, the effect is the same as raising the maximum zero fuel weight of the airplane. For dispatch with center wing tank fuel override pumps inoperative, any center fuel contained within these tanks must be considered as part of zero fuel weight.

10.12 Wing Loads for Structural Analysis

Consideration will be given to the resulting net loads applied to the analysis of the wing box structure. If the wing stress analysis is based on beam theory at a section normal to the wing box, which has been the traditional method of analysis before the introduction of finite element methods, then the net loads are summed to obtain shear, moment, and torsion along a preselected load reference axis.

For stress analysis of the wing structure using finite element analysis methods, the resulting aerodynamic loads and the wing internal loads due to inertia must be distributed in a preselected manner on the structural model of the wing.

10.12.1 Wing Load Reference Axis

The wing load reference axis (LRA) is the spanwise locus of reference points at each of the wing stations that have been selected for stress analysis of the wing box structure as shown in Fig. 10.1. This axis is fixed for a given aircraft structural configuration and is assumed not to vary with load condition.

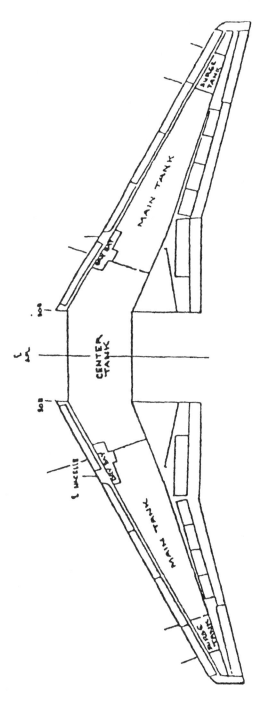

Fig. 10.20   Wing fuel tanks selected for bending moment optimization.

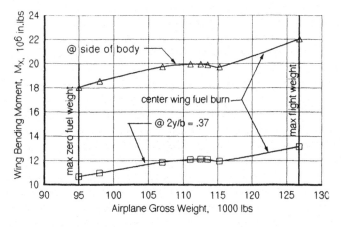

Fig. 10.21   Wing-bending moment vs airplane gross weight effect of fuel usage.

10.12.2 Wing Elastic Axis

The elastic axis is usually defined as the locus of points at which normal loads (V_z, M_x, and T_{lra} in Fig. 10.1) can be applied without causing the wing to twist.[6] In essence, the elastic axis would be drawn through the shear centers of each structural section chosen for stress analysis of the wing box structure. In reality, the shear center of a given box structure will vary depending on the type of loading applied and whether cutouts or significant discontinuities are designed into the structure, such as landing gear beams or wing-mounted nacelles.

For practical purposes the wing elastic axis will be selected to represent the center of twist at each wing section such that a common axis may be assumed for all conditions. For swept-back wings this axis may sweep aft of the center between the front and rear spars as the axis approaches the side of the body. This is usually based on test data that show the rear spar may be carrying proportionately higher loads than the front spar.

For practical purposes the elastic axis and the load reference axis are assumed the same. In essence, what goes into an aeroelastic load analysis as an elastic axis comes out as the load reference axis.

10.12.3 Wing Beam Shear, Moment, and Torsion

Wing net beam shear, bending moment, and torsion along the wing load reference axis, as shown in Figs. 10.2, 10.3, and 10.4, are calculated as the summation of the net airloads and inertia loads outboard of the analysis stations.

The spanwise distributions of aerodynamic loads are usually integrated with respect to freestream axes as shown in Fig. 10.22. The shear, moment, and torsion about the load reference axis due to airload are then calculated from freestream loads using Eqs. (10.27–10.29):

$$M_x = M \cos \tau - T \sin \tau - (\blacktriangle M_x)_a + (\blacktriangle M_x)_b \qquad (10.27)$$

$$T_{lra} = T \cos \tau + M \sin \tau - (\blacktriangle T_{lra})_a + \blacktriangle T_{lrab} \qquad (10.28)$$

$$V_z = V - (\blacktriangle V_z)_a + (\blacktriangle V_z)_b \qquad (10.29)$$

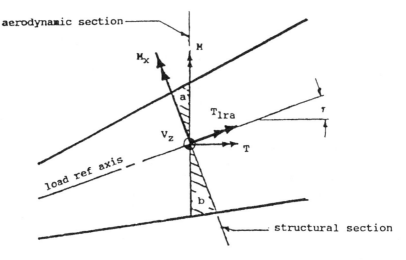

Fig. 10.22 Rotational corrections for wing beam loads.

where V, M, and T are the integrated aerodynamic loads reference to the freestream axis (lb and in.-lb); and $\blacktriangle V_z$, $\blacktriangle M_x$, and $\blacktriangle T_{lra}$ are the incremental shear, moment, and torsion due to the aerodynamic loading of sections a and b, shown in Fig. 10.22 (lb and in.-lb).

The incremental loads on panels a and b of Fig. 10.22 are obtained by integration of the pressures acting over these panels for the specific condition under investigation.

10.12.4 Wing Chordwise Shear, Moment, and Axial Load

Chordwise loads are obtained from integrated wind-tunnel pressure data referenced to the wing section chord plane. The spanwise distributions of chord shear and moment, V_x and M_z shown in Fig. 10.1, include the effect of wing twist and deflection.

The relationship of section lift and drag to chord force is shown in Eq. (10.30):

$$C_c = C_d \cos\alpha - C_l \sin\alpha \qquad (10.30)$$

where C_c is the chordwise force coefficient, C_l is the section lift coefficient, C_d is the section drag coefficient, and α is the section angle of attack (deg).

If an estimation of the wing section drag is available, then the chordwise force acting on a given wing section may be calculated from Eq. (10.30). If pressure distributions are obtained from wind-tunnel data, the chordwise forces acting on the wing may be obtained directly by integration with respect to the selected chord plane at each analysis wing section.

Chordwise shear and bending moment and axial loads are calculated from the chord forces, accounting for wing deflection and twist if a flexible wing analysis is used.

10.13 Simplified Shear Flow Calculations for Spars

The criticality of a given condition cannot be determined using the shear and torsion envelopes as shown in Figs. 10.2 and 10.4. Since shear and torsion usually are related to the wing spar design conditions, a simplified approach may be used to identify the critical conditions for wing spars and related structure.

Neglecting spar shear flow induced by chordwise shear and bending, one can write the equations for front and rear spar shear flow as shown in Eqs. (10.31) and (10.32). If the wing box structure has a midspar, a similar equation, shown in Eq. (10.31), may be written, although shear flow may not be of significance for the criticality of this spar.

The shear flow in the front and rear spars are written as functions of the wing shear, moment, and torsion at a given structural analysis station using the sign convention shown in Fig. 10.1:

$$Q_{fs} = a_1 V_z + b_1 M_x + c_1 T_{lra} \qquad (10.31)$$

$$Q_{rs} = a_2 V_z + b_2 M_x + c_2 T_{lra} \qquad (10.32)$$

where Q is the spar shear flow (lb/in.), V_z is the shear normal to the wing reference plane (lb), M_x is the beam-bending moment (in.-lb), and T_{lra} is the torsion about the load reference axis (in.-lb).

The coefficients shown in Eqs. (10.32) and (10.33) may be obtained from unit load solutions run through the wing box stress analysis. In actuality since the shear flows calculated from these two equations are only used to assess one condition relative to another, the cofficients a_i and c_i could be obtained from the relationship of the front and rear spar locations from the load reference axis at each load station. The effect of induced shear flow due to beam bending cannot be obtained in this manner.

A set of shear flow coefficients used for wing analysis load surveys is shown in Table 10.3 for a typical commercial jet transport.

Table 10.3   Wing spar shear flow calculations using the simplified approach; the shear flow coefficients are defined by Eqs. (10.31) and (10.32) for the front and rear spars[a]

2y/b	Front spar			Rear spar		
	a_1	b_1	c_1	a_2	b_2	c_2
0.90	72.9	−163	2391	−61.6	175	2391
0.80	64.2	−138	1643	−55.2	138	1643
0.73	57.7	−120	1302	−52.9	129	1302
0.63	52.3	−95	997	−46.7	99	997
0.53	46.9	−83	791	−42.9	83	791
0.45	44.4	−74	661	−40.3	76	661
0.35	40.9	−132	513	−35.4	124	513
0.28	33.1	−211	379	−28.9	195	379
0.20	22.4	−193	222	−21.6	190	222

[a]The data shown are only representative of the complete set of coefficients required for an adequate load survey to select critical wing design conditions. The coefficients have the following scale factors applied to shear, moment, and torsion: shear: $10^{-3} V_z$ (lb), moment: $10^{-6} M_x$ (in.-lb), and torsion: $10^{-6} T_{lra}$ (in.-lb).

10.14 Wing Spanwise Load Distributions

Consideration will be given to several methods for obtaining the spanwise airload distribution over a wing for both rigid and flexible analyses.

If the wing spanwise lift distributions are obtained from wind-tunnel pressure data, the analysis for a rigid wing may be readily accomplished. Wind-tunnel data may be integrated for the lift and pitching moment variation with angle of attack and Mach number.

If airplane tail-off aerodynamic data, but not pressure data, are available, or if an aeroelastic analysis is desired, then the methods discussed in the following sections may be used to calculate the spanwise distribution of loads on a straight or swept wing.

10.14.1 Method of NACA Report 921

One of the simplest methods for obtaining the spanwise lift distribution for symmetrical flight load analysis is presented in Ref. 1. This method, with some restrictions, is applicable for analysis of wings of arbitrary planform.

The theory used for this analysis is based on the work of Weissinger as summarized in Ref. 2.

The analysis describes a set of seven equations representing the relationship between wing angle of attack at each station and the resulting spanwise load distribution. This set reduces to four equations per side for a symmetrical load analysis, since the distributions are the same on each wing. The lift at the airplane centerline is common for both wings. This relationship is shown in Eq. (10.33) using matrix notation:

$$[a]\{G\} = \{\alpha\} \qquad (10.33)$$

where a_{ij} is the aerodynamic coefficient indicating the influence of the spanwise lift at station j on the downwash angle at span station i, $\{G\}$ is the dimensionless circulation $\{\Gamma/bv\} = \{C_l c/2b\}$, $\{\alpha\}$ is the angle of attack (rad), b is the wing span (ft), c is the wing section streamwise chord (ft), C_l is the section lift coefficient, v is the freestream velocity (ft/s), and Γ is the circulation (ft²/s).

The influence matrix $[a]$ is determined as a function of Mach number. This is sometimes called the "planform distortion method." For the traditional lifting line subsonic theory, the center of lift of each aerodynamic panel is assumed at the quarter-chord of the section. In this method the center of lift is allowed to vary with Mach number, but the downwash angle is still measured at the center of lift plus one-half the section chord. For the traditional approach this would be at the three-quarters chord location.

Although Ref. 1 does not specifically discuss an aeroelastic analysis, the method is readily adapted to include the influence of an elastic wing by introducing the change in angle of attack due to airload. By adding the lift and pitching moment equations as is done in Ref. 4, one can derive a closed solution.

A similar method of analysis for calculating the antisymmetrical load distribution for rolling maneuvers is shown in Ref. 3.

10.14.2 Method of NACA TN 3030

A method for computing the steady-state span load distribution on an airplane wing of arbitrary planform and stiffness is presented in Ref. 4. The analysis as developed in this report is applicable to both symmetrical and antisymmetrical flight maneuver conditions.

The symmetrical analysis includes a set of equations representing the lift and pitching moments of the total airplane, such that a closed solution of the resulting system of equations is possible. These equations are solved for the spanwise lift distribution, wing angle of attack, and balancing tail load for a specific gross weight, load factor, and center of gravity position.

The antisymmetrical analysis is solved for the spanwise load distribution due to rolling velocity. The aeroelastic spanwise load distribution due to roll acceleration may also be calculated for an elastic wing.

The aerodynamic influence matrix $[S_1]$ as derived in Ref. 4 is applicable to a flat twisted wing and does not account for out-of-plane surfaces such as winglets. Wing flaps and control surfaces such as ailerons and spoilers may be included in this analysis. External store airloads are included in terms of the contribution to wing aeroelastic loading.

A method of reducing wind-tunnel data based on integrated wing section pressure distributions is discussed in Appendix G of Ref. 4. Flight load surveys made on several commercial jet transport configurations have shown a good correlation of measured wing loads to analytical loads using the analysis methods of this report.

References 5 and 6 are included as sources of some of the other methods published by NACA on static aeroelastic analyses of swept and unswept wings.

10.14.3 Doublet-Lattice Method

The doublet-lattice method may be used for interacting lifting surfaces in subsonic flow. The theory and methods are beyond the scope of this book but are presented in Refs. 7 and 8.

The theoretical basis of the doublet-lattice method is linearized aerodynamic theory. The undisturbed flow is uniform and is steady for maneuver conditions or unsteady for gust analyses.

The principle advantage of the doublet-lattice method is the ability to analyze nonplanar configurations such as winglets placed at the tips of the wing and to provide nodal loads for finite element analyses.

References

[1]DeYoung, J., and Harper, C. W., "Theoretical Symmetric Span Loading at Subsonic Speeds for Wings Having Arbitrary Plan Form," NACA Rept. 921, 1948.

[2]Weissinger, J., "The Lift Distribution of Swept-Back Wings," NACA TM 1120, 1947.

[3]DeYoung, J., "Theoretical Antisymmetrical Span Loading for Wings of Arbitrary Plan Form at Subsonic Speeds," NACA Rept. 1056, 1951.

[4]Gray, W. L., and Schenk, K. M., "A Method for Calculating the Subsonic Steady-State Loading on an Airplane with a Wing of Arbitrary Plan Form and Stiffness," NACA TN 3030, Dec. 1953.

[5]Diederich, F., "Calculation of the Aerodynamic Loading of Swept and Unswept Flexible Wings of Arbitrary Stiffness," NACA Rept. 1000, 1950.

[6]Diederich, F. W., and Foss, K. A., "Charts and Approximate Formulas for the Estimation of Aeroelastic Effects on the Loading of Swept and Unswept Wings," NACA Rept. 1140, 1953.

[7]Rodden, W. P., Giesing, J. P., and Kalman, T. P., "Refinement of the Nonplanar Aspects of the Subsonic Doublet-Lattice Lifting Surface Method," *Journal of Aircraft,* Vol. 9, No. 1, 1972.

[8]Giesing, J. P., Kalman, T. P., and Rodden, W. P., "Subsonic Unsteady Aerodynamics for General Configurations, Part I—Vol. I—Direct Application of the Nonplanar Doublet Lattice Method," Air Force Flight Dynamics Lab., AFFDL-TR-71-5, Wright-Patterson AFB, OH, Nov. 1971.

11
Body Monocoque Loads

Body monocoque loads, although fairly simple to calculate, have evolved over the years since the early commercial jet transports to the present series of aircraft. The methodology has changed primarily due to the increased capability of digital computers to handle large amounts data.

Before 1970, monocoque loads were analyzed using stress analysis beam theory to calculate bending stresses and shear flow at a given body station. Since the inception of finite element analysis methods, structural loads analyses have been modified to accommodate these advanced techniques.

11.1 Monocoque Analysis Criteria

The criteria for flight maneuvers, gust conditions, and landing and ground-handling loads are the same for the monocoque analysis as for the horizontal tail, vertical tail, and wing structure. Cabin pressure is combined with flight and landing conditions as discussed in Sec. 11.6.

The use of rational loading conditions has been allowed by the certifying agencies to meet one of the needs of the sophisticated analytical tools used for stress analysis of the monocoque structure. When stress analyses were accomplished using simple beam methods, load envelope conditions were used where each analysis station was analyzed using the maximum loads at a selected station without concern that the conditions could be different at the adjacent fore and aft stations.

Finite analysis tools require that the system being analyzed has a set of balanced loads, such that all of the loads coming from the wing, empennage, and landing gears are in equilibrium. Rational loading conditions allow the engineer to meet the requirement for a set of balanced loads when finite element models are used for structural analysis of the fuselage monocoque.

11.2 Monocoque Design Conditions

The determinations of body monocoque loads for static load conditions are readily obtained using the sum of the loads from the wing, empennage, and landing gears and the airload and inertia loads acting on the monocoque structure.

Body monocoque load envelopes are shown in Figs. 11.1 and 11.2 for vertical design conditions and Figs. 11.3–11.5 for lateral design conditions for a typical jet transport.

In addition to the conditions shown in these figures, dynamic loads acting on the monocoque structure must be determined for the flight gust conditions discussed in Chapter 5, the dynamic landing loads discussed in Sec. 10.8, and the dynamic taxi analysis discussed in Sec. 10.9.

11.2.1 Body Airload

Body airloads are calculated from integrated pressure data obtained from wind-tunnel tests as a function of Mach number, angle of attack, and sideslip angle.

191

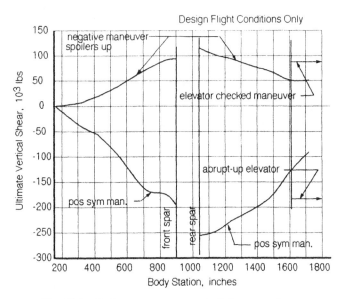

Fig. 11.1   Body monocoque vertical shear envelope.

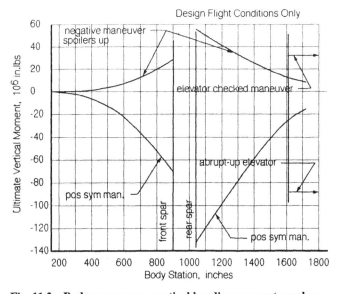

Fig. 11.2   Body monocoque vertical bending moment envelope.

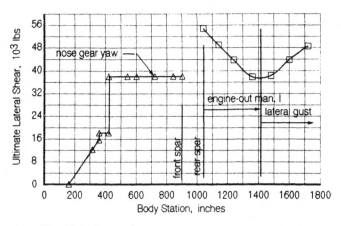

Fig. 11.3   Body monocoque lateral shear envelope.

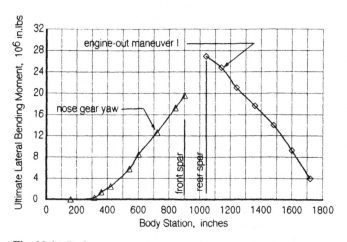

Fig. 11.4   Body monocoque lateral bending moment envelope.

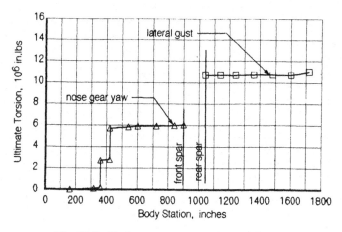

Fig. 11.5   Body monocoque torsion envelope.

Body airload is a significant factor in calculating net body loads for symmetrical flight conditions and contributes to a reduction in monocoque loads.

In general, if body airload is neglected in the calculation of monocoque loads, the airplane would be aerodynamically out of balance, and hence the stress engineer would have difficulty in accomplishing a finite element analysis of the monocoque structure.

For symmetrical flight conditions, the local body pressures, shown in Fig. 11.6, are integrated around the body circumference at a given station to obtain the local lift as a function of dynamic pressure, angle of attack, and Mach number. Lateral load conditions are accomplished in a similar manner. Local pressures are integrated to give the side load as a function of dynamic pressure, sideslip angle, and Mach number.

11.2.2 Flight Load Conditions

The flight loads applied to the body monocoque are from the horizontal tail, vertical tail, wing, and engines that may be externally mounted on the fuselage.

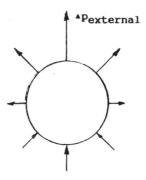

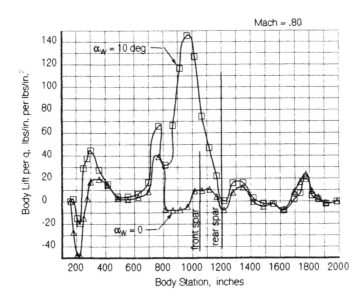

Fig. 11.6 Body airload distribution.

Because the fuselage is a lifting surface, the monocoque structure is subjected to an airload acting over the length of the body as discussed in Sec. 11.2.1.

The flight conditions for the horizontal tail, vertical tail, and wing loads are discussed in other chapters of this book. The flight criteria discussed in Chapters 2–5 are applicable to monocoque loads.

An assessment of the body monocoque loads must be made for the vertical and lateral gust conditions discussed in Chapter 5. The effects of structural dynamics must be considered for both conditions. In general, the forward body becomes critical for vertical gust conditions and the aft body for lateral gust conditions.

11.2.3 Unsymmetrical Flight Load Considerations

Special consideration needs to be given to the lateral gust or yawing maneuver conditions in terms of the level flight condition that is combined with lateral loads. The effect on the aft body of the combined loads is shown schematically in Fig. 11.7.

For conditions where the horizontal tail loads are negative (downward-acting loads), producing tension in the upper aft body crown and compression in the lower aft body monocoque, the combination with the lateral loads will be additive in quadrants in one and three. For conditions where the horizontal tail loads are positive (upward-acting loads), the opposite is true; the loads are additive in quadrants in two and four.

For example, the downward-acting horizontal tail load may be for a forward center of gravity position, whereas the upward-acting horizontal tail load could be for an aft center of gravity position with speedbrakes extended.

Depending on the symmetrical nature of the structure for lateral conditions, the analyst must consider the lateral loads reversed in direction (but not magnitude) from the direction shown in Fig. 11.7. This becomes important when one side of the monocoque has a different structural configuration, such as a body door cutout.

One of the unsymmetrical conditions that may become critical for the aft body monocoque is the oblique gusts discussed in Sec. 5.10. Particular attention must be given to how these loads are calculated, such that structural weight is not added to the aircraft.

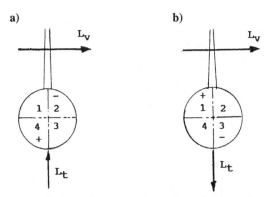

Fig. 11.7   Unsymmetrical loading conditions: a) up horizontal tail load condition and b) down horizontal tail load condition. + structure in tension; − structure in compression.

11.2.4 Landing and Ground-Handling Load Conditions

The landing and ground-handling load conditions that are applied to the monocoque structure are discussed in Chapters 6 and 7 along with the applicable criteria.

Dynamic landing analyses must be accomplished, as discussed in Secs. 6.9 and 10.8, to determine the resulting loads on the body monocoque structure. These loads may be critical on the vertical bending structure aft of the wing body rear spar bulkhead, depending on the criticality of the flight maneuver conditions.

In a similar manner, dynamic taxi analyses must be determined, as discussed in Secs. 7.3.2 and 10.9, to obtain the resulting body loads. This condition is usually not critical for the monocoque but may be of concern for the nose gear support structure when taxied over very rough runways.

11.3 Load Factors Acting on the Body

The load factors acting along the body must be determined using the methods discussed in Chapter 5 for vertical and lateral gust conditions. In general, these will be greater than the maneuver load factors calculated using analyses discussed in Chapters 2–4.

The vertical gust load factor for a given airspeed will vary with airplane gross weight as shown in Fig. 11.8. The load factors shown in this figure were calculated using the gust load formula in Eq. (5.4) for positive gust velocities as shown in Table 5.2.

The variation of load factors along the body monocoque are shown in Fig. 11.9. During the certification of jet transports in the early 1960s, the certifying agencies became concerned with some flight test data measured on a B-47 bomber that indicated an increase in load factor at the nose of the airplane as the aircraft penetrated a vertical gust. This event was considered to be caused by aircraft pitch-up as the gust was encountered, not due to dynamic response of the structure. At that time analytical methods of an aircraft penetrating a discrete gust were somewhat lacking.

To provide a conservatism in calculating the vertical load factors acting on the forward body during a gust encounter, the decision was made to neglect the pitching acceleration relief as noted in Fig. 11.9. This conservatism was discarded in

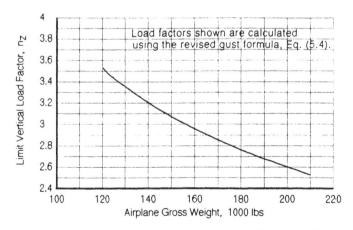

Fig. 11.8 Vertical gust load factor variation with gross weight.

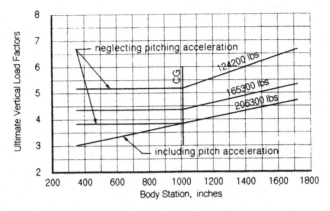

Fig. 11.9   Vertical gust load factor variation along body.

later aircraft designs as more sophisticated mathematical tools were made available that properly accounted for aerodynamic parameters and structural dynamic considerations.

A comparison is shown in Fig. 11.10 of the gust load factors calculated using the gust formula, Eqs. (5.4), (5.12), and (5.13), with a dynamic analysis using the proposed discrete gust requirements discussed in Sec. 5.3.1. The load factors calculated using the gust formula are based on the gust velocities shown in Table 5.2 and Fig. 5.2. The example shown in Fig. 11.10 indicates a good agreement of the load factors at the airplane center of gravity, but the obvious effect of body flexibility is evident by the load factors at the extreme ends of the body.

11.3.1   Maximum Vertical Load Factors

The design requirements of FAR 25.561, Emergency Landing Conditions, must be considered in determining the load factors to be applied to equipment, cargo, or other such items within the monocoque cabin where occupants must be protected. Amendment 25-64 in 1988 revised FAR 25.56 to incorporate higher crash load factors as shown in Table 11.1.

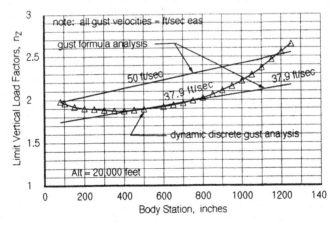

Fig. 11.10   Comparison of load factors from dynamic analysis with gust formula analysis.

Table 11.1   Emergency landing load factors
per FAR 25.561

	Ultimate load factors	
	Before 1988	Amendment 25-64
Upward	2.0	3.0
Downward	4.5	6.0
Forward	9.0	9.0
Sideward	±1.5	±3.0[a]
Rearward	None	1.5

[a]4.0 on seats and attachments.

Load factors for equipment support designed within the passenger cabin or cockpit would become critical for the emergency landing conditions shown in Table 11.1, when compared with the load factors shown in Fig. 11.9, depending on the certification date of the aircraft. The example in Fig. 11.9 indicates that the load factors due to a vertical gust are critical in the extreme aft end of the body, even when compared with the requirements per Amendment 25-64.

11.4   Payload Distribution for Monocoque Analysis

One of the most significant parameters that affects the magnitude of the body monocoque loads is the distribution of payload within the body, whether passengers, cargo, or both.

How this payload is distributed, and the concentration at any one point in the body, must be given consideration with respect to the impact on the monocoque structural requirements. This also includes the requirements for the floor beams necessary to support the passenger seats, galleys, or cargo.

11.4.1   Historical Perspective Using Couple Loads

Before certification of the 707-100 in September 1958, body monocoque loads were calculated using a technique developed on previous propeller aircraft whereby a minimum number of body panel loads were obtained from the weights engineer. These panel loads consisted of the following three conditions.

1) The payload aft of the rear spar bulkhead was end loaded at the maximum payload rate up to the rear spar bulkhead. The remaining payload to obtain maximum zero fuel weight was loaded in the forward body such that the most forward center of gravity possible was obtained. The resulting airplane center of gravity was always aft of the design forward center of gravity limit.

2) The payload forward of the front spar bulkhead was end loaded at the maximum payload rate up to the front spar bulkhead. The remaining payload to obtain maximum zero fuel weight was loaded aft of the rear spar bulkhead starting from the rear loading point. The airplane center of gravity for this condition was then obtained.

3) The monocoque panel loads for the empty airplane were also provided along with the resulting center of gravity position.

Couple loads were then applied to the appropriate panel load condition to correct the actual center of gravity position to the design forward or aft position as the case warranted.

For example, if the desired flight condition was a balance maneuver at the airplane design forward center of gravity, then couple loads defined by Eqs. (11.1) and (11.2) were determined as necessary to transfer the center of gravity for the maximum aft body panel condition to the forward limit.

Assume couple loads were applied as shown in Fig. 11.11 to the two selected body stations. The couple loads were arbitrarily applied at body stations forward and aft of the possible cargo or passenger loading:

$$P_f = -P_a \tag{11.1}$$

The transfer equation may be written as follows:

$$W B S_{cg1} + P_a B S_a - P_f B S_f = W B S_{cg} \tag{11.2}$$

Combining Eqs. (11.1) and (11.2), one can determine the magnitude of the couple load:

$$P_a = W(B S_{cg1} - B S_{cg})/(B S_a - B S_f) \tag{11.3}$$

where W is the maximum zero fuel weight (lb), $B S_{cg1}$ is the airplane center of gravity for actual payload condition (in.), $B S_{cg}$ is the desired analysis center of gravity position (in.), $B S_a$ is the aft couple load location (in.), and $B S_f$ is the forward couple load location (in.).

The use of couple loads is shown in Table 11.2 for an assumed airplane. The loads shown are calculated for payload only for simplicity of analysis.

To obtain loads for an aft center of gravity condition, one would calculate the couple loads in a similar manner, but they would have signs opposite to those of the forward center of gravity condition shown.

11.4.2 Rational Loading Conditions

With the advent of modern digital computers, the use of the couple load technique was eliminated in favor of rational loading conditions. The rational loading methodology was accepted by the U.S. Civil Aeronautics Authority (CAA) [predecessor to the Federal Aviation Administration (FAA)] during the certification of the 707-100 jet transport in September 1958.

An example of a rational condition that maximizes the aft body bending condition with the airplane loaded at the forward center of gravity limit is shown in Fig. 11.12. As shown by comparing the loads calculated in Tables 11.2 and 11.3, shear and bending moments are reduced by using rational loading conditions vs the couple load method. The most significant change is in shear.

The main purpose of eliminating the couple load method was to reduce the conservative body monocoque loads, hence a reduction in structural weight. The basic problem of using rational loading of the body payload is the large increase in number of monocoque load conditions necessary for structural analysis.

A summary of design payload conditions used for a typical freighter airplane is shown in Fig. 11.13, where the rational loading conditions are shown on a design center of gravity envelope along with the couple load example shown in Table 11.3.

Various aft body conditions are investigated to maximize bending moments aft of a given monocoque section using end-loaded conditions as shown in Fig. 11.12.

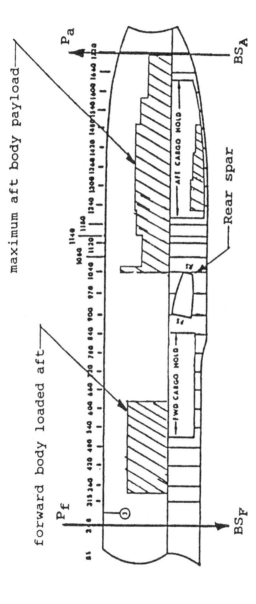

Fig. 11.11 Body monocoque loading using couple loads; airplane condition: design forward center of gravity at maximum zero fuel weight. Notes: 1) aft body payload is loaded at the maximum capability aft of the wing rear spar bulkhead station, and 2) forward body is loaded aft at the maximum rate of loading to obtain forward payload defined as forward payload = maximum payload – aft body payload; maximum payload = MZFW – OEW.

Table 11.2   Body monocoque loads using couple loads to correct to forward
center of gravity condition; payload condition: aft body loaded
to maximum payload as defined in Fig. 11.11

	Body station	Weight, lb	BS_{cg}	CG	Shear, lb	Moment, 10^6 in.-lb
Aft body payload						
	1,040 RS	1,681	1,048.5	—	—	—
	1,057	5,500	1,101.5	—	—	—
	1,146	6,200	1,190.5	—	—	—
	1,235	7,400	1,279.5	—	—	—
	1,324	7,100	1,368.5	—	—	—
	1,413	6,900	1,457.5	—	—	—
	1,502	5,800	1,546.5	—	—	—
	1,591	6,500	1,653.5	—	—	—
	1,716	0	0	—	—	—
Total airplane data						
Operating empty						
weight	1,016.90	113,862	—	0.125	—	—
Aft cargo	1,362.77	47,081[a]	—	—	—	—
Forward cargo	616.99	39,057	—	—	—	—
Max. zero fuel						
weight actual	1,020.22	200,000	—	0.142	—	—
Max. zero fuel						
weight desired	1,009.87[b]	200,000	—	0.090[b]	—	—
Couple loads required [see Eq. (11.3)]						
Body station	1,720	248	—	—	—	—
Load, lb	1,406	−1,406	—	—	—	—
Aft body shear and bending moment						
at $n_z = 1.0$ (payload only)						
Without couple load	1,040	—	—	—	−47,081	−15.196
With couple load	1,040	—	—	—	−45,675	−14.240

[a]Maximum aft payload.
[b]Design forward center of gravity.

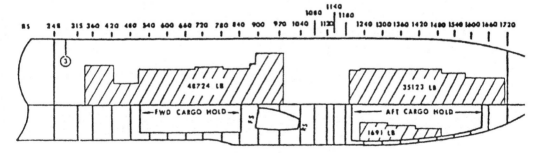

Fig. 11.12   Body monocoque rational loading. The airplane is loaded to maximum
zero fuel weight at the most forward flight center of gravity position; the payload is
extreme-end loaded to maximize aft body bending moment.

Table 11.3   Body monocoque loads using rational payload for forward center of gravity condition; payload condition: aft body loaded to maximum payload for forward center of gravity as defined in Fig. 11.12

	Body station	Weight, lb	BS_{cg}	CG	Shear, lb	Moment, 10^6 in.-lb
Aft body payload						
	1,040 RS	0	—	—	—	—
	1,057	0	—	—	—	—
	1,146	0	—	—	—	—
	1,235	6,200	1,286.72	—	—	—
	1,324	7,100	1,368.5	—	—	—
	1,413	6,900	1,457.5	—	—	—
	1,502	5,800	1,546.5	—	—	—
	1,591	6,500	1,653.5	—	—	—
	1,716	0	0	—	—	—
Total airplane data						
OEW	1,016.90	113,862	—	0.125	—	—
Aft cargo	1,460.56	32,500	—	—	—	—
Forward cargo	721.88	53,638	—	—	—	—
MZFW actual	1,009.87	200,000	—	0.090	—	—
MZFW desired	1,009.87[b]	200,000	—	0.090[b]	—	—
Aft body shear and bending moment at $n_z = 1.0$ (payload only)						
Rational condition	1,040	—	—	—	−32,500	−13.668
Couple condition[a]	1,040	—	—	—	−45,675	−14.240

[a]See Table 11.2.
[b]Design forward center of gravity.

Aft body shear conditions at a given body station may be maximized by using reverse loading conditions where payload is loaded aft of a specific station.

The two examples shown in this chapter are only a part of the many conditions that require investigation for structural analysis of the body monocoque structure and related structure within the airframe. Forward body monocoque loads may be maximized using a similar procedure as that for the aft body. Gust conditions must be investigated for the combination of payload that will give the design condition for a given set of floor beams.

11.5   Monocoque Payload Limitations

Monocoque payload limitations used in the airplane weight and balance manuals, which are a part of the FAA certification requirements, are shown in Fig. 11.14 for a cargo airplane. Monocoque payload limitations for passenger airplanes are developed for the airplane weight and balance manuals in a similar manner. These limitations are used to determine the monocoque loading distribution for structural analyses.

For cargo aircraft, an additional graph is usually provided in the form of what has become known as the payload shear curve, as shown in Fig. 11.15. This curve is the envelope of the maximum payload, at a load factor of 1.0 g, for which

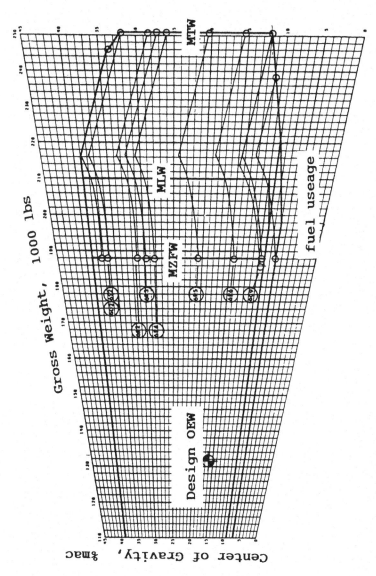

Fig. 11.13 Design center of gravity envelope for monocoque conditions; the conditions shown are loaded in a rational manner to maximize monocoque loads. (Numbers in circles indicate conditions.)

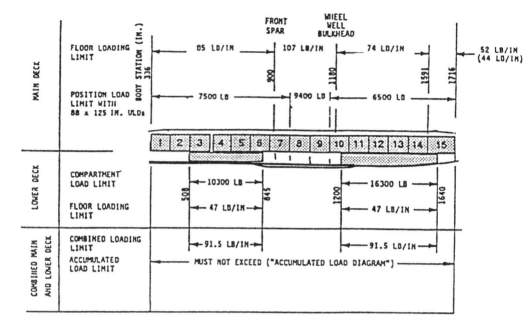

Fig. 11.14   Cargo loading structural limitations typical of freighter-type aircraft. The structural limitations shown are depicted in the airplane Weight and Balance Manual.

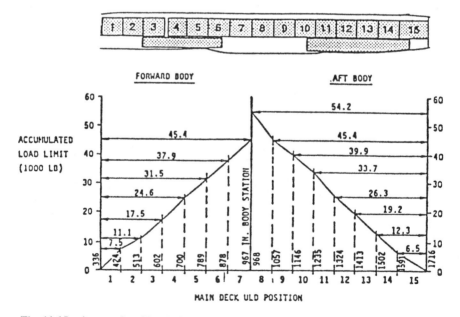

Fig. 11.15   Accumulated load diagram structural limitations; the curve shown is also called a "payload shear curve." The accumulated load diagram is depicted in the airplane Weight and Balance Manual.

Table 11.4 Cabin pressure criteria for typical jet transport[a]

	Design limit pressure, lb/in.^{2}	Factor	Design ultimate pressure, lb/in.^{2}	FAR reference
Pressure only	9.10	2.0	18.2	25.365(d)
Pressure combined with flight loads	9.10	1.5	13.65 plus external pressure	25.365(a)
Pressure combined with landing loads	0.333[b]	1.5	0.50	25.365(c)

[a]The cabin pressure system has the following settings: maximum relief valve setting = 8.95 ± 0.15 lb/in.^{2} and maximum operating pressure differential = 8.6 ± 0.10 lb/in.^{2}
[b]Maximum pressure allowed for landing for this aircraft.

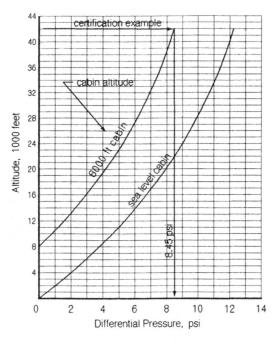

Fig. 11.16 Cabin differential pressure for aircraft certified to 42,000 ft maximum altitude.

the monocoque is designed. By loading the cargo within the monocoque limits shown, the operators of these freighters are only constrained by maximum zero fuel weight and the forward and aft center of gravity limitations of the airplane.

11.6 Cabin Pressure Criteria

The cabin pressure criteria, as stated in FAR 25.365, are shown in Table 11.4 for a typical jet transport certified in the 1980s. As noted, three sets of pressures are used for design: 1) pressure only, where the monocoque structure is designed without the addition of flight or landing loads; 2) pressure combined with flight loads, which includes the external pressure distribution for the condition under consideration; and 3) pressure combined with landing loads.

The variation of cabin pressure with altitude is shown in Fig. 11.16 as applied to an airplane that has a maximum certified altitude limit of 42,000 ft. The cabin pressure system is designed to maintain cabin altitude of 8000 ft or less throughout the airplane flight envelope.

12
Control Surface Loads and High-Lift Devices

The purpose of this chapter is to provide an overview of the determination of control surface and high-lift device loads. Consideration will be given to any special criteria applicable to the determination of the loads for these surfaces.

The control surfaces considered in this chapter are 1) ailerons, 2) elevators, 3) rudders, 4) spoilers on the wing, and 5) tabs. Movable horizontal stabilizers, although a control surface, are discussed in Chapter 8.

The wing high-lift devices considered in this chapter are 1) leading-edge Krueger flaps, 2) leading-edge slats, and 3) trailing-edge flaps.

12.1 Control Surface Loads

The basic premise for determining control surface loads is that the control surface hinge moments are available from wind-tunnel tests or flight tests on aircraft with similar configurations. For very large surfaces, pressure distributions may be available from wind-tunnel tests.

12.1.1 Design Criteria

The design criteria per FAR 25.391 and 25.393 state that the control surfaces must be designed for the limit loads resulting from the following conditions: 1) pitch maneuver flight conditions per FAR 25.331, 2) rolling maneuver flight conditions per FAR 25.349, 3) yawing maneuver flight conditions per FAR 25.351, 4) unsymmetrical loads per FAR 25.427, 5) ground gust conditions per FAR 25.415, and 6) loads parallel to the hinge line per FAR 25.393.

In determining control surface loads, the effect of pilot effort, trim tabs, anti-balance tabs, and power control units (PCU) must be considered as defined in FAR 25.397, 25.407, and 25.409.

12.1.2 Control Surface Hinge Moments

Control surface hinge moments are usually obtained from wind-tunnel tests and verified by flight tests. In the commercial jet transports such as the 747, 757, and 767 aircraft where the primary control surfaces (ailerons, elevators, and rudder) are actuated by PCUs, the maximum hinge moment available is directly obtainable from the PCU output.

For configurations involving tabs and aerodynamic balance panels designed between 1953 and 1970, the method of analysis for the calculation of pressure distributions becomes more complex.

A good source of hinge moment and chordwise pressure distribution data is Ref. 1, when other sources of wind-tunnel data are not available. This was used on the early vintage commercial jets for initial design; then the final hinge moments were verified by flight tests.

The total hinge moment about the surface hinge line may be considered to be made up of the sum of the various components of the configuration as follows:

$$HM_{input} = HM_{aerodynamic} \qquad (12.1)$$

where HM_{input} is the hinge moment required from pilot effort plus PCU,

$$HM_{input} = HM_{pe} + HM_{pcu} \qquad (12.2)$$

$$HM_{aerodynamic} = HM_{cs} + HM_{tabs} + HM_{bal\ panels} \qquad (12.3)$$

where HM_{cs} is the hinge moment about the control surface hinge line due to control surface aerodynamic loading (ahead of and behind the hinge line), HM_{tab} is the hinge moment about the control surface hinge line due to balance or antibalance tab aerodynamic loading, and $HM_{bal\ panels}$ is the hinge moment about the control surface hinge line due to internal aerodynamic balance panels.

Each of the increments defined in Eq. (12.2) will be considered in determining the hinge moment required to balance the aerodynamic hinge moment for the control surface position desired. The resulting load distribution due to the aerodynamic increments defined by Eq. (12.3) will be discussed independently.

12.1.3 Hinge Moment from Pilot Effort and PCUs

The control system must be designed to provide the required hinge moment about the control surface hinge line to produce the desired surface motion. Before the inclusion of PCUs, the amount of hinge moment available was directly from pilot effort. The design requirements for pilot effort were limited as prescribed in FAR 25.397(b), 300 lb for elevator and rudders and $80D$ in.-lb for ailerons with control system configurations using wheels instead of a stick. The term D is the control wheel diameter in inches per FAR 25.397(c).

With the advent of PCUs, control system designs evolved with most if not all of the input hinge moment coming from the PCU.

An example of the hinge moment available for a rudder control system is shown in Fig. 12.1, in which the primary input to the rudder is from a PCU; the pilot effort contributes only a small input in the power-on mode. For this airplane the tabs revert to a balance mode to assist the pilot in obtaining the level of surface motion required in the failure condition.

12.2 Determination of Maximum Available Control Surface Angle

An example of the determination of maximum elevator available as a function of airspeed is shown in Fig. 12.2. Aerodynamic hinge moment coefficients are shown as a function of elevator angle and Mach number.

By cross-plotting the hinge moment available from the PCU plus pilot effort with the aerodynamic hinge moment coefficients as shown in Fig. 12.2, one can determine the maximum elevator angle as a function of airspeed and Mach number. The hinge moment available is determined in coefficient form using Eq. (12.5).

The hinge moment about the control surface hinge line is defined in terms of the hinge moment coefficient, dynamic pressure, and reference area and chord:

$$HM_{cs} = CH_{cs}q(Sc)_{cs} \qquad (12.4)$$

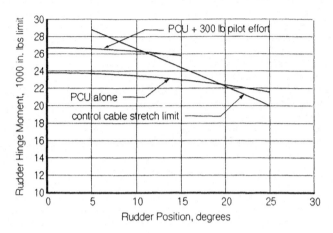

Fig. 12.1 Hinge moment available from power control unit plus pilot effort; hinge moments are for a rudder system, and the stretch limit is with the rudder pedals bottomed.

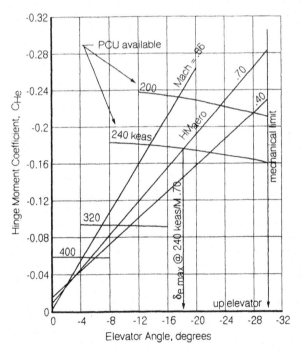

Fig. 12.2 Determination of maximum elevator available vs airspeed and Mach number.

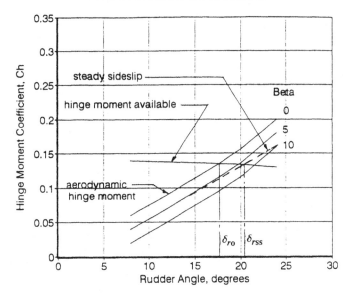

Fig. 12.3 Graphical solution for rudder available in steady sideslip; the rudder data assume rudder blowback during the maneuver. δ_{ro} = rudder available at zero sideslip; δ_{rss} = rudder available in a steady sideslip.

Solving for the hinge moment coefficient representing the power available to the control surface,

$$CH_{cs} = HM_{cs}/[q(Sc)_{cs}] \qquad (12.5)$$

where HM_{cs} is the hinge moment available from the PCU plus pilot effort (ft-lb), q is the dynamic pressure (lb/ft^2), and $(Sc)_{cs}$ is the aerodynamic reference area and chord for the control surface (ft^3).

The effects of angle of attack or sideslip angle in determining the aerodynamic hinge moment should be considered. An example is the effect of determining the rudder available in a steady sideslip as shown in Fig. 12.3.

Examples of the maximum rudder available for two types of rudder systems are shown in Figs. 4.3 and 4.4. The first figure is shown for a system with a two-stage pressure reducer that activates at a given airspeed. The second figure is shown for a system whereby the rudder is limited by a ratio changer that varies with airspeed. The ratio changer alters the effective moment arm of the PCU such that the rudder angle decreases with increasing airspeed, while still providing the pilot with full rudder pedal available at all airspeeds.

12.3 Control Surface Airload Distribution

For control surfaces in which the pressure distributions are not available from wind-tunnel or flight tests, the following procedures have been used. The assumption is made that the hinge moment about the control surface hinge line is known.

Examples of the variation of hinge moment coefficient due to control surface deflection are shown in Figs. 12.2 and 12.3 as a function of Mach number and control surface position.

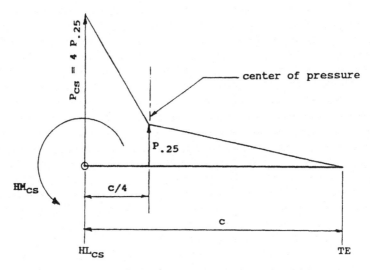

Fig. 12.4   Control surface chordwise pressures assuming a distribution with center of pressure at 0.25c.

12.3.1   Chordwise Pressures with Center of Pressure at 0.25c

Chordwise and spanwise pressure distributions may be calculated for control surfaces by assuming a shape such as the distribution shown in Fig. 12.4, which has the center of pressure at the quarter-chord of the control surface. Furthermore, the spanwise load distribution is assumed to vary as a function of the control surface chord.

By taking the moment about the hinge line, one can derive the relationship between chordwise pressures and hinge moment:

$$P_{0.25avg} = 4HM_{cs}/(Sc)_{cs} \qquad (12.6)$$

The spanwise pressure distributions are determined from Eqs. (12.7) and (12.8):

$$P_{0.25cs} = P_{0.25avg}(c/c_{cs}) \qquad (12.7)$$

$$P_{cs} = 4P_{0.25cs} \qquad (12.8)$$

where c is the control surface chord (in.), and c_{cs} is the control surface reference chord used in hinge moment coefficient calculations, $(Sc)_{cs}$ (in.).

Of the two chordwise distributions discussed in this section and Sec. 12.3.2, the condition whereby the center of pressure is assumed at the quarter-chord of the control surface will give the higher total airload over the surface for the same input hinge moment. This will represent chordwise pressures for low Mach conditions as can be seen from the distributions shown in Ref. 1.

12.3.2   Chordwise Pressures with a Triangular Distribution

A second distribution may be assumed for conditions where the chordwise pressures are assumed triangular as shown in Fig. 12.5. For this analysis the center of pressure is at the 33% chord, and the spanwise load distribution varies as a function of control surface chord.

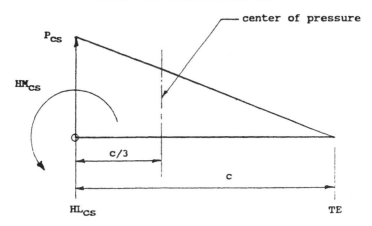

Fig. 12.5   Control surface chordwise pressures assuming a triangular distribution.

By taking the moment about the hinge line, one can derive the relationship between chordwise pressures and hinge moment:

$$P_{cs\ avg} = 6HM_{cs}/(Sc)_{cs} \tag{12.9}$$

The spanwise pressure distributions are determined from Eq. (12.10), using the average pressures calculated from Eq. (12.9):

$$P_{cs} = P_{cs\ avg}(c/c_{cs}) \tag{12.10}$$

Higher Mach number conditions may be represented by the triangular airload distribution, although some conditions may be more representative by using a trapezoidal distribution. This type of distribution may be used to provide an aft loaded condition that may be used for design of the control surface trailing-edge structure. In all cases, the further aft the chordwise center of pressure is, the lower the total airload to produce the same hinge moment.

12.3.3   Incremental Airload Distribution Due to Tabs

The airload distribution due to control surface tabs may be calculated in a similar manner to the main control surface by assuming a triangular variation of airload as shown in Fig. 12.6. This distribution is assumed to be effective over the area as shown in the figure.

The tab hinge line pressure may then be calculated knowing the control surface hinge moment due to tab as shown in Eq. (12.11):

$$P_t = 6HM_{tab}/[2(Sc)_{tf} + S_{ta}(3c_{tf} + c_{ta})] \tag{12.11}$$

where HM_{tab} is the hinge moment due to the tab about the control surface hinge line (ft-lb), $(Sc)_{tf}$ is the effective area and chord forward of the tab hinge line (ft³), and $(Sc)_{ta}$ is the effective area and chord aft of the tab hinge line (ft³).

Hinge moment coefficients, referenced to the elevator hinge line for an elevator–tab configuration, are shown in Fig. 12.7.

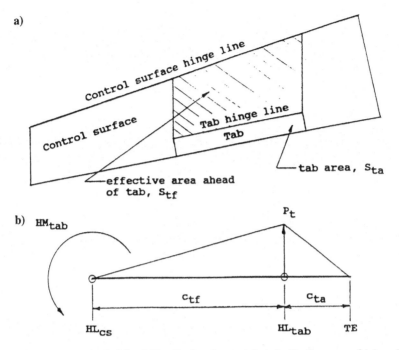

Fig. 12.6   Incremental airload distribution due to tabs: a) effective area of tab and b) chordwise airload distribution.

12.4   Tab Design Airload

Tab design loads may be determined based on the maximum design tab hinge moment about the tab hinge line and assuming a chordwise distribution whereby the center of pressure is at the quarter-chord of the tab, similar to Fig. 12.4. The spanwise load distribution varies as a function of the chord:

$$P_{t0.25\text{avg}} = 4HM_{\text{tab}}/(Sc)_{\text{tab}} \tag{12.12}$$

$$P_{t\text{hl}} = 4P_{t0.25\text{avg}} \tag{12.13}$$

where HM_{tab} is the tab hinge moment about the tab hinge line, and $P_{t\text{hl}}$ is the tab hinge line pressure (see Fig. 12.4).

Tab hinge moment coefficients may be obtained from Ref. 1 or other sources such as wind-tunnel or flight tests.

12.5   Spoiler Load Distribution

The spoiler load distributions may be obtained directly from the hinge moment capability of the spoiler actuators as shown in Fig. 12.8. For the in-flight conditions with the spoilers extended, two distributions are assumed, each producing the same hinge moment as defined by the extension capability of the spoiler actuators. The condition whereby P_1 at the spoiler leading edge is defined by Eq. (12.14) produces the largest airload of the two, which will become the critical design condition for the spoiler hinges and related structure.

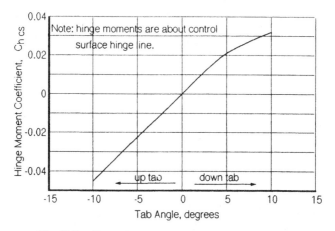

Fig. 12.7   Control surface hinge moment due to tab.

The relationship between chordwise pressure and spoiler hinge moment is defined in Eq. (12.14):

$$P_1 = 6HM_{\text{ext}}/(Sc)_{\text{sp}} \qquad (12.14)$$

The aft loaded condition will design the spoiler trailing-edge structure and has been selected to provide adequate structure to withstand buffeting that may occur at maximum spoiler extension. The relationship between P_1 and P_2 becomes

$$P_2 = 0.5P_1 \qquad (12.15)$$

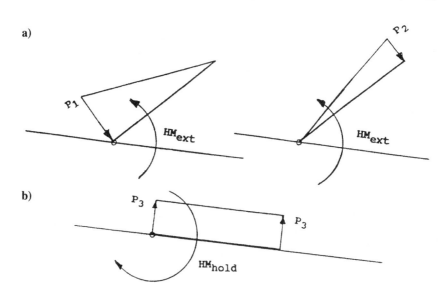

Fig. 12.8   Spoiler load distribution: a) maximum airload spoilers extended and b) maximum airload spoilers retracted.

Table 12.1 Example of spoiler airload distribution[a] with flight spoilers: $(Sc)_{sp} = 31.64$ ft^3 and ground spoilers: $(Sc)_{sp} = 21.21$ ft^3

Condition	Spoiler position, deg	System pressure, lb/in.^{2}	HM, ft-lb limit	lb/in.^{2} limit		
				P_1	P_2	P_3
Flight spoilers						
Blowdown	17.5	3000	2890	5.71	2.85	———
Hold down	0	3900	2600	———	———	1.71
Ground spoilers						
Blowdown	55	———	1000[b]	2.95	1.47	———
Hold down	0	3900	2400	———	———	2.36

[a]The airload distributions acting on the spoilers for extended and retracted conditions are shown in Fig. 8. Chordwise pressures are calculated using Eqs. (12.14–12.16).
[b]Maximum aerodynamic hinge moment with spoiler extended on the ground.

The spoiler hold-down condition must be considered in design of the hold-down mechanism and related spoiler structure. Assuming a distribution of pressure as shown in Fig. 12.8, one can calculate the chordwise pressures from Eq. (12.16):

$$P_3 = 2HM_{hold}/(Sc)_{sp} \qquad (12.16)$$

The parameters shown in Eqs. (12.14–12.16) are defined as follows: HM_{ext} is the hinge moment capability of the spoiler actuator for extension or blowdown (ft-lb), HM_{hold} is the hinge moment capability of the spoiler actuator to hold the spoiler in the closed or down position (ft-lb), and $(Sc)_{sp}$ is the spoiler reference area and chord (ft^3).

The load distributions for a typical spoiler panel are calculated in Table 12.1. Note that design loads are calculated for in-flight spoilers that are used as speed-brakes and roll control devices and ground spoilers that are activated to dump lift from the wing and flaps during landing roll-out. These spoilers are normally activated by switches in the landing gear oleo system.

12.6 Structural Deformation of Control Surface Hinge Lines

Because of the nature of the design of control surfaces with multiple hinge points such as ailerons, elevators, rudders, and tabs, consideration must be given to the redistribution of hinge point loads due to structural deformation of the surface to which they are attached.

For example, elevator hinge loads will vary considerably due to elevator hinge line deformation induced by the aeroelastic characteristics of the stabilizer. Hence, for two conditions that have the same elevator angle and pressure distribution, the hinge loads will vary significantly depending on the load distribution on the horizontal tail stabilizer. Although the summation of all of the hinge loads is not altered, the loads will redistribute due to structural flexibility.

In the early days before the introduction of finite element analyses, control surface hinge point loads due to airload and the redistribution of load due to structural flexibility were computed (initially by hand, then later by computers) using the moment distribution or Hardy Cross method introduced in 1932 by Cross (see Refs. 2 and 3).

Current aircraft include these effects in the finite element analysis models to determine control surface hinge loads by defining, for example, the stabilizer and elevator as part of the total structural model. Similar models are defined for wings with multiple hinge ailerons or the vertical tail and rudders.

12.7 High-Lift Devices

The wings of commercial jets from the early aircraft certified in 1956 to the current models have been equipped with various high-lift devices to enhance low-speed approach and flight characteristics. The design load characteristics of the following high-lift devices are considered: 1) leading-edge Krueger flaps—used on early 707/720 and 737 models to provide a two-position leading-edge flap (extended or retracted) that enhances the flow over the wing in the high angle-of-attack flight regime with trailing-edge flaps partially or fully extended; 2) leading-edge slats—used on the 727 and later versions of the 737, 757, and 767 models, to provide a two- or three-position slat that enhances the flow over the wing in the high angle-of-attack flight regime at all trailing-edge flap detents; and 3) trailing-edge flaps—occur in various configurations from the triple-slotted flaps on the 727 and 737 models to the double-slotted flaps on the 707, 757, and 767 aircraft. These flaps are used to enhance low-speed approach, thus allowing lower landing speeds, hence shorter field lengths as required for landing.

12.7.1 Krueger Flaps

Leading-edge Krueger flaps, shown in Fig. 12.9, are critical for positive loads at the upper left-hand corner of the V-n diagram and negative loads at the lower right-hand corner, as shown in Fig. 12.10.

Positive load conditions are critical at the design maneuver load factor in the flaps-down condition, $n_z = 2.0$, at the maximum lift coefficient for the flap detent position under consideration. Two flap detent positions are shown in Fig. 12.11 whereby the Krueger flap normal force coefficients are shown as a function of tail-off lift coefficient. These coefficients are related to the overall lift balance for the airplane using Eq. (12.17):

$$C_{Lto} = (n_z W - L_t)/(q S_w) \tag{12.17}$$

where n_z is the airplane normal load factor, and L_t is the balancing tail load for the condition under investigation (lb).

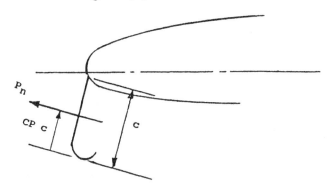

Fig. 12.9   Leading-edge Krueger flap.

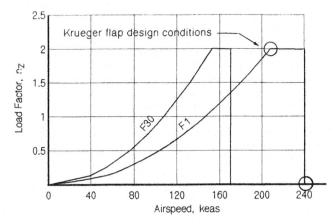

Fig. 12.10   Krueger flap design conditions.

Krueger flap design loads are shown in Table 12.2. The critical positive load condition is defined by the lower of the two flap detent positions shown for the example airplane. It should also be noticed that for a constant flap detent position Krueger flap loads are less with decreasing lift coefficient even though the dynamic pressure q is increasing.

The negative load condition is usually defined by the $n_z = 0$ load factor at the placard speed V_F as shown in Fig. 12.10 and in the calculations for the example shown in Table 12.2.

12.7.2 Leading-Edge Slats

Leading-edge slats are shown in Fig. 12.12 for an airplane with a three-position configuration: fully extended for landing flap detents, partially extended with essentially zero trailing-edge gap for takeoff flap detents, and fully retracted. Other configurations may have only two slat positions, extended and retracted.

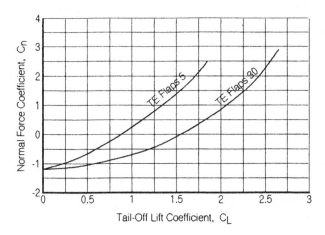

Fig. 12.11   Krueger flap normal force coefficients.

Table 12.2   Example of Krueger flap design loads calculated at one spanwise station using the load coefficients shown in Fig. 12.11[a]

	At $C_{N\max}$	Effect of increasing airspeed	At $C_{N\max}$	Effect of increasing airspeed	At V_F speed
n_z	2.0	2.0	2.0	2.0	0
Flaps	30	30	5	5	5
$C_{L\text{tail}-\text{off}}$	2.70	2.50	1.85	1.50	0
q, lb/in.^{2}	0.514	0.556	0.751	0.926	1.471
V_e, keas	147.8	153.6	178.6	198.3	250
Krueger flap loads, limit					
C_n	3.00	2.30	2.50	1.40	−1.20
P_n, lb/in.	27.8	23.0	33.8	23.3	−31.8
CP, %c/100	0.37	0.39	0.37	0.42	0.44

[a]Conditions: $W = 100,000$ lb, $S_w = 1000$ ft^2, $V_F = 250$ keas at flaps 5, and $c = 18$ in. For this analysis we assume that $L_t = 0$ in Eq. (12.17). The term $c =$ Krueger flap reference chord at the analysis station.

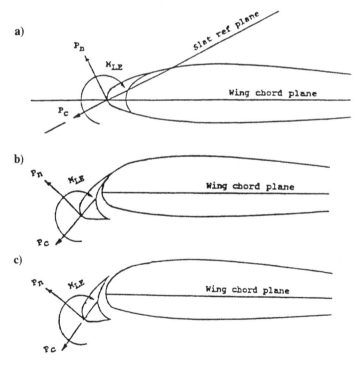

Fig. 12.12   Three-position leading-edge slat: a) wing slat retracted, b) wing slat partially extended, and c) wing slat fully extended.

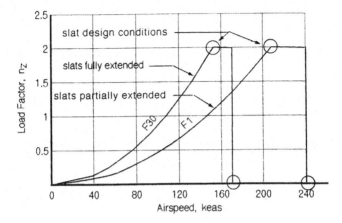

Fig. 12.13 Slat design conditions—flaps down.

Leading-edge slats are critical for positive loads at the upper left-hand corner of the V-n diagram and negative loads at the lower right-hand corner, as shown in Fig. 12.13 for flaps-down conditions.

Positive load conditions are critical in a manner similar to Krueger flaps, i.e., at the design maneuver load factor in the flaps-down configuration, $n_z = 2.0$, at the maximum lift coefficient for the flap detent position under consideration. The variation of slat load coefficients with reference angle of attack is shown in Fig. 12.14.

Leading-edge slat design loads for a typical slat are shown in Table 12.3. The critical positive load occurs at the upper left-hand corner of the V-n diagram at the maximum lift coefficient for the flap detent position under consideration. The maximum negative load occurs at $n_z = 0$ at the placard speed V_F.

As noted in Fig. 12.15, the leading-edge slats may be critical at the upper left-hand corner of the flaps-up V-n diagram for slat bending induced by wing flexibility. Wing bending induces reactions in the slat structure, which, along with

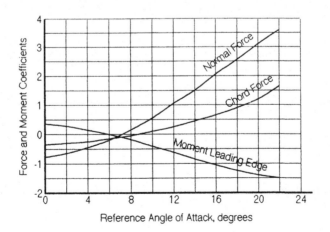

Fig. 12.14 Slat load coefficients.

Table 12.3 Example of leading-edge slat design loads calculated at one spanwise station using the load coefficients shown in Fig. 12.14[a]

	At C_{Nmax}	Effect of increasing airspeed		At V_F speed
n_z	2.0	2.0	2.0	0
Flaps	20	20	20	20
C_L	2.22	2.10	1.94	0
α_{ref}, deg	16.0	14.0	12.0	0.22
q, lb/in.^{2}	0.626	0.661	0.716	1.245
V_e, keas	163.0	167.6	174.4	230
Slat loads, limits				
C_n	2.10	1.60	1.10	−0.75
P_n, lb/in.	27.8	22.3	16.6	−19.7
C_c	0.60	0.40	0.25	−0.40
P_c, lb/in.	7.9	5.6	3.8	−10.5
C_{mLE}	−1.05	−0.85	−0.65	0.45
M_{LE}, in.-lb/in.	−293.5	−250.8	−207.8	250.1

[a]Conditions: $W = 100,000$ lb, $S_w = 1000$ ft^2, $V_F = 230$ keas at flaps 20, and $c = 21.13$ in. The term $c =$ slat reference chord at the analysis station.

the external pressure distribution acting on the slat upper surface, are then reacted by the slat actuators.

12.7.3 Trailing-Edge Flaps

An example of a trailing-edge double-slotted flap configuration is shown in Fig. 12.16. Critical design conditions are shown in Fig. 12.17 for two flap detent positions.

Comparison of the positive maneuver loads at the placard speed V_F will show that these loads are less than the head-on gust condition. The load factor for the head-on gust condition is calculated using Eq. (5.52) as derived in Chapter 5.

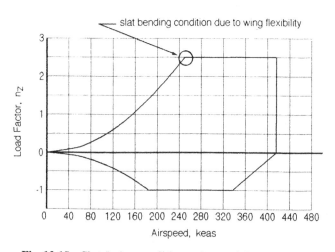

Fig. 12.15 Slat design condition—slats and flaps retracted.

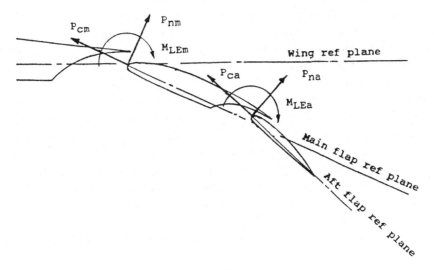

Fig. 12.16   Trailing-edge double-slotted flap. P_n = normal force flap, lb/in.; P_c = chord force flap, lb/in.; M_{LE} = pitching moment about flap leading edge, in.lb/in.; a = aft flap; and m = main flap.

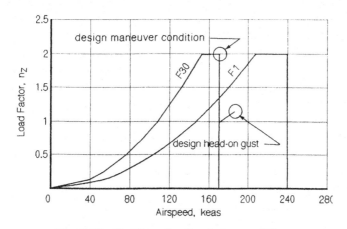

Fig. 12.17   Trailing-edge flap design conditions.

12.7.4   Effect of Spoilers on Trailing-Edge Flap Loads

The trailing-edge flaps may be affected by the presence of spoilers acting as speedbrakes or for roll control in the flaps-down configuration. The load distribution on the flap immediately behind the spoilers may show an added increment of load and must be accounted for in the analysis.

If, for example, spoilers are used for roll control, then the condition may become more critical at the symmetrical maneuver load factor for roll maneuver at placard speed than the head-on gust condition. This phenomenon is configuration dependent and should be given close scrutiny during the design of the flap structure.

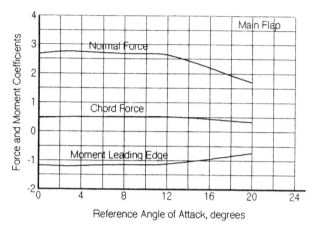

Fig. 12.18   Trailing-edge main flap load coefficients.

References

[1]Tinling, B. E., and Dickson, J. K., "Tests of a Model Horizontal Tail of Aspect Ratio 4.5 in the Ames 12-Foot Pressure Wind Tunnel. I—Quarter-Chord Line Swept Back 35°," NACA RM A9G13, Sept. 1949.

[2]Bruhn, E. F., *Analysis and Design of Aircraft Structures,* Purdue Univ. Press, Lafayette, IN, Jan. 1949.

[3]Perry, D. J., *Aircraft Structures,* McGraw–Hill, New York, 1949.

13
Static Aeroelastic Considerations

The term "aeroelastic" is applied to a class of phenomena that involves the interaction of aerodynamic, inertial, and elastic forces in a structure.[1] The interaction of these forces can give rise to a variety of aeroelastic phenomena, such as 1) flutter and static divergence (aeroelastic instabilities); 2) transient or dynamic responses as a result of external forces such as vibrations, buffeting, gusts, taxi over rough runways, and landing impact; 3) control reversal; and 4) reduction in aircraft flight control characteristics and redistribution of loads due to structural flexibility.

Aeroelasticity is concerned with stiffness, not strength, and the interacting effect on related aerodynamic loading of aircraft flight surfaces, wings, and empennage.[2]

The discussion in this chapter will be limited to static divergence, control reversal, and aircraft flight control characteristics.

13.1 Flutter, Deformation, and Fail-Safe Criteria

Flutter, deformation, and fail-safe criteria as stated in FAR/JAR 25.629 include the requirements concerning static aeroelastic phenomena. The subject of flutter, although not covered in this text, does involve the interaction of aerodynamic loads, inertia loads, and structural deformation and is of major concern in the design of aircraft structure. The airspeed margins required for flutter, static deformation prevention, and control reversal are the same.

Amendment 25-77 to FAR Part 25, issued June 22, 1992, defines the aeroelastic stability requirements[1] as follows:

FAR 25.629(a) *General*: The aeroelastic stability evaluations required under this section include flutter, divergence, control reversal and undue loss of stability and control as a result of structural deformation. . . .

FAR 25.629(b) *Aeroelastic Stability Envelopes*: The airplane must be designed to be free from aeroelastic instability for all configurations and design conditions within the aeroelastic stability envelopes as follows:

(1) For normal conditions without failures, malfunctions, or adverse conditions, all combinations of altitudes and speeds encompassed by the V_D/M_D versus altitude envelope enlarged at all points by an increase of 15 percent in equivalent airspeed at both constant Mach number and constant altitude. In addition, a proper margin of stability must exist at all speeds up to V_D/M_D and there must be no large and rapid reduction in stability as V_D/M_D is approached. The enlarged envelope may be limited to Mach 1.0 when M_D is less than 1.0 at all design altitudes.

13.1.1 Historical Perspective

Before June 22, 1992, the flutter and divergence margin was 20% above V_D/M_D for commercial aircraft. This margin was originally based on the concept that the ratio of dynamic pressure at flutter speeds to the dynamic pressure at V_D/M_D

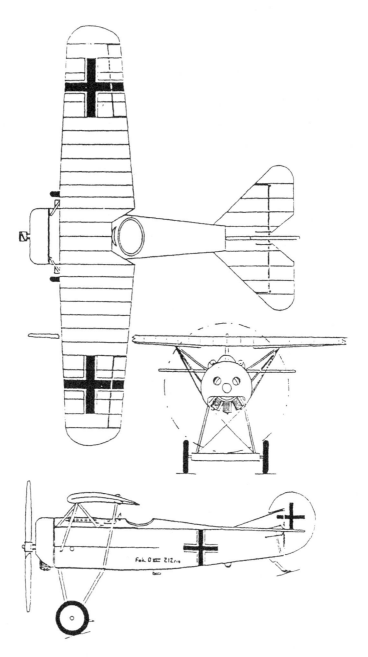

Fig. 13.1   Example of early airplane design with static divergence problem (Fokker D-8, vintage 1918).

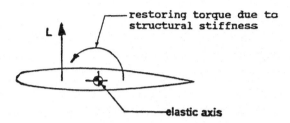

Fig. 13.2   Static aeroelastic divergence. (Static aeroelastic divergence will occur when the lift on the wing or tail about the elastic axis produces a torque that overcomes the restoring forces due to structural stiffness.)

should be at least 1.5; hence, the velocity ratio would be $\sqrt{1.5} = 1.22$ (Ref. 2). This is analogous to the factor of safety applied to limit load to obtain ultimate design load.

13.2   Static Divergence Analysis

One of the most significant studies done on an early airplane that had static divergence problems was the Fokker D-8 shown in Fig. 13.1, designed during World War I for the German Army by Anthony Fokker. During dives in combat some of the early production airplane wings failed, causing loss of aircraft and pilot. German Army experts static tested the wings and proved the structure to be adequate to carry the design loads required for combat aircraft. During further static tests by Fokker, he observed that the wing diverged with application of increasing load; thus he concluded that the wings had failed due to static divergence.[2]

Static aeroelastic divergence will occur at the airspeed where the lift on the wing or tail produces a torque about the elastic axis that will overcome the restoring forces due to structural stiffness, as shown in Fig. 13.2.

In general, for aircraft with swept-back wings or empennages more than 4–8 deg, static divergence does not exist. This may be seen in the chart shown in Fig. 13.3 as obtained from Ref. 3. For swept-back surfaces of more than 4–8 deg, the chart indicates that the dynamic pressure parameter is negative; thus the static divergence does not exist for these surfaces. This was the conclusion that was made during the early certification programs of the 707, 727, and 737 aircraft, making use of the studies and statements summarized in Ref. 4.

Appendix H of Ref. 5 presents a method to calculate the dynamic pressure for wings that may have divergence concerns. In general, this would be for straight or swept-forward wings. As noted in Ref. 5, it is conceivable that with a large external store, such as a tip tank, the wing could have a divergence problem.

13.3   Control Surface Reversal Analysis

Control surface reversal analyses must be considered for swept-back surfaces due to the nature of the interaction of aerodynamic forces produced by the control surfaces and structural deformation. The control surface aerodynamic effectivity is reduced by aeroelasticity and may even reverse as airspeeds increase.

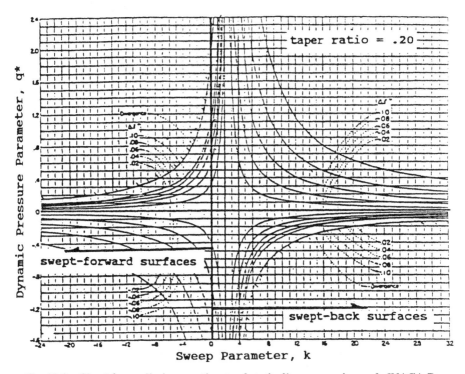

Fig. 13.3   Chart for preliminary estimate of static divergence airspeeds [NACA Report 1140 (Ref. 3)]. q^* = function of dynamic pressure, q; k = dimensionless sweep parameter (see reference).

Control surface reversal will occur when the aerodynamic lift produced by the control surface is overcome by the aerodynamic loading due to aeroelastic effects induced by wing bending and torsion as shown in Fig. 13.4.

Solution of Eq. (13.1) will give the span lift distribution due to control surface deflection $\{lcs\}$, which may be integrated to determine the control surface effective lift:

$$[(1/4q)[1/m_0][Sa] - [Se]]\{lcs\} = \{\alpha_{CS}\} \qquad (13.1)$$

where q is the dynamic pressure (lb/in.^{2}), m_0 is the two-dimensional lift curve slope per radian, $[Sa]$ is the aerodynamic induction matrix, $[Se]$ is the structural elasticity matrix, and $\{\alpha_{CS}\}$ is the section angle of attack due to control surface angle (rad).

Using the procedure developed in Ref. 5, or other similar methods, the aeroelastic effects of control surfaces such as ailerons, spoilers, elevators, and rudders may be determined. The aerodynamic induction matrix $[Sa]$, as shown in Eq. (13.1), may be calculated using the symmetrical or antisymmetrical analysis depending on the control surface being investigated: 1) ailerons—analysis based on antisymmetrical $[Sa]$, 2) spoilers—analysis based on antisymmetrical $[Sa]$ when spoilers are used for lateral control, 3) elevators—analysis based on symmetrical $[Sa]$, and 4) rudders—analysis based on symmetrical $[Sa]$ if the assumption is made that the horizontal tail acts as an end plate.

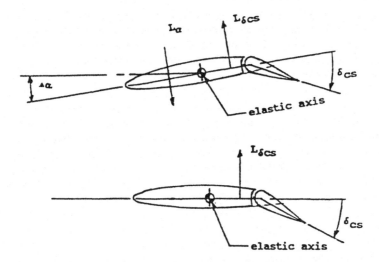

Fig. 13.4 Control surface reversal due to aeroelasticity. (Reversal will occur when the aerodynamic lift produced by the control surface is overcome by the aerodynamic loading due to aeroelasticity.) L_α = lift due to change in wing angle of attack; $L_{\delta CS}$ = lift due to control surface; δ_{CS} = control surface angle; $\triangle\alpha$ = change in angle of attack due to bending and torsion.

The section aerodynamic coefficients representing the control surface are usually based on linearized data and may be represented as follows:

$$\{\alpha_{CS}\} = \{\alpha_1\}\left(\frac{d\alpha}{d\delta}\right)\delta_{CS} \tag{13.2}$$

$$\{Cm_{CS}\} = \{Cm_1\}\left(\frac{dCm}{d\delta}\right)\delta_{CS} \tag{13.3}$$

where δ_{CS} is the control surface position, and $\{\alpha_1\}$ and $\{Cm_1\}$ are the normalized section coefficients.

Equations (13.2) and (13.3) defining the control surface effectiveness parameters, $d\alpha/d\delta$ and $dCm/d\delta$, are shown in a form that allows investigation of the importance of these parameters on control surface reversal speeds.

Section aerodynamic parameters are usually obtained as a function of Mach number from wind-tunnel data. These parameters, represented by Eqs. (13.2) and (13.3), are normalized as required to obtain $\{\alpha_1\}$ and $\{Cm_1\}$. The resulting normalized parameters may vary with Mach number or may be coalesced for all Mach numbers. Normalized coefficients are shown in Table 13.1.

The calculations of the aileron reversal speeds are shown in Table 13.2 with the results summarized in Fig. 13.5. The primary parameter contributing to control surface reversal is section pitching moment. The relationship of $dCm/d\delta$ to $d\alpha/d\delta$ is important and must be given attention when these parameters are calculated from integrated pressure data obtained from the wind tunnel. Depending upon the stiffness characteristics of the surface to which the aileron, elevator, or rudder

Table 13.1 Typical set of normalized section aerodynamic coefficients as represented in Eqs. (13.2) and (13.3) calculated for an aircraft with an outboard aileron; Mach number effects are considered in the selection of $(d\alpha/d\delta)^a$ and $(dCm/d\delta)^b$ shown in Eqs. (13.2) and (13.3)[c, d]

Station 2y/b	Section data at $M = 0.40$		Normalized data at $M = 0$	
	$\{\alpha\}$, rad/deg	$\{Cm\}$, /deg	$\{\alpha_1\}$, rad/deg	$\{Cm_1\}$, /deg
0.95	0.00336	−0.00253	0.00840	0.666
0.85	0.00698	−0.00380	0.01745	1.000
0.75	0.00151	−0.00165	0.00378	0.434
0.55	0	0	0	0

[a]$d\alpha/d\delta = 0.400$.
[b]$dCm/d\delta = -0.0038$/deg.
[c]The assumption for these distributions is that the normalized variation spanwise does not vary with Mach number, only the coefficients $d\alpha/d\delta$ and $dCm/d\delta$.
[d]The variation spanwise exists because the ailerons do not fully encompass the aerodynamic panels shown inboard and outboard of $2y/b = 0.85$.

is attached, an increase of 20% on the section pitching moment may reduce the reversal speed as much as 50 kn equivalent airspeed.

Examples are shown in Figs. 13.6–13.8 for the substantiation of reversal speeds requirements of FAR 25.629 for an aircraft certified before June 1992, which had higher speed margin requirements based on $1.2V_D$.

For lateral control configurations using both ailerons and spoilers, aeroelastic characteristics are assessed using the combination as shown in Fig. 13.7.

13.4 Structural Stiffness Considerations

Consideration must be given to the stiffness parameters used in the aeroelastic analyses for determining structural loads, reversal analyses, and aeroelastic effects on stability and control flight parameters.

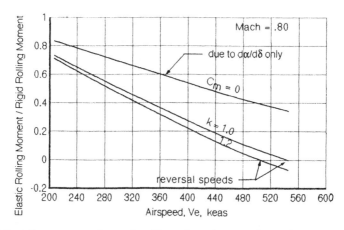

Fig. 13.5 Aileron reversal speeds, effect of section aerodynamic parameters; $k =$ increase in $dCm/d\delta$ (see Table 13.2).

Table 13.2 Aileron reversal analysis (summarized in Fig. 13.5)—effect of section aerodynamic parameters on reversal speeds using the following baseline data: Mach = 0.80, dα/dδ = 0.235, dCm/dδ = –0.0042/deg, and ratio = (dCm/dδ)/(dα/dδ) = –0.01787

		Due to dα/dδ = 1			Due to dCm/dδ = 1	Net
q, psi	V_e, keas	RM_e[a]	Rm_r[b]	RM_e/RM_r $cm = 0$	$\blacktriangle RM_e$[a]	RM_e/RM_r[c]
			Baseline analysis			
1	206.1	0.5526	0.6625	0.834	3.829	0.731
3	357.0	1.2026	1.9876	0.605	26.080	0.371
5	460.9	1.4954	3.3127	0.451	56.758	0.145
7	545.3	1.5864	4.6378	0.342	89.481	−0.003
		Increase ratio of dCm/dδ *to* dα/dδ *by k* = 1.10				
1	206.1	0.5526	0.6625	0.834	3.829	0.720
3	357.0	1.2026	1.9876	0.605	26.080	0.347
5	460.9	1.4954	3.3127	0.451	56.758	0.115
7	545.3	1.5864	4.6378	0.342	89.481	−0.037
		Increase ratio of dCm/dδ *to* dα/dδ *by k* = 1.20				
1	206.1	0.5526	0.6625	0.834	3.829	0.710
3	357.0	1.2026	1.9876	0.605	26.080	0.324
5	460.9	1.4954	3.3127	0.451	56.758	0.084
7	545.3	1.5864	4.6378	0.342	89.481	−0.072

[a]RM_e = rolling moment elastic, 10^6 in.-lb.
[b]RM_r = rolling moment rigid, 10^6 in.-lb.
[c]Net $(RM_e/RM_r) = (RM_e/RM_r)_{cm=0}$ k $\Delta(RM_e)(RM_r)$ where k = increase in dCm/dδ.

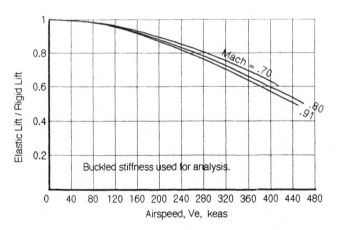

Fig. 13.6 Effect of aeroelasticity on pitch control due to elevator. The end point of each curve is $1.2V_D$ at constant Mach number for the aircraft used in this example; the speed margin required at the time of certification of this aircraft exceeds the margin of 1.15 adopted June 1992.

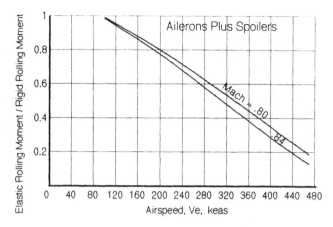

Fig. 13.7 Effect of aeroelasticity on lateral control due to aileron plus spoilers. The end point of each curve is $1.2V_D$ at constant Mach number for the aircraft used in this example; the speed margin required at the time of certification of this aircraft exceeds the margin of 1.15 adopted June 1992.

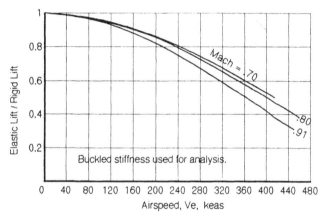

Fig. 13.8 Effect of aeroelasticity on directional control due to rudder. The end point of each curve is $1.2V_D$ at constant Mach number for the aircraft used in this example; the speed margin required at the time of certification of this aircraft exceeds the margin of 1.15 adopted June 1992.

As structure is loaded during flight, the stiffness may vary according to whether the structure is buckled or unbuckled. The following are general guidelines that are recommended for design.

Structural load analyses: Stiffness parameters are usually obtained for buckled surfaces. At design limit loads the structural skins are usually buckled under compression loads.

Reversal analyses: Stiffness parameters are usually based on buckled surfaces that will give lower reversal speeds and will be conservative.

Flight control aeroelastic characteristics: Aeroelastic analyses may be calculated using both buckled and unbuckled stiffness. Data for high-speed dive recovery should be calculated using buckled stiffness.

References

[1]Anon., "Vibration, Buffet and Aeroelastic Stability Requirements for Transport Category Airplanes," Amendment No. 25-77 to FAR Part 25, Federal Aviation Regulations, U.S. Dept. of Transportation, June, 1992.

[2]Collar, A. R., "The First Fifty Years of Aeroelasticity," *Aerospace Magazine,* Feb. 1978.

[3]Diederich, F. W., and Foss, K. A., "Charts and Approximate Formulas for the Estimation of Aeroelastic Effects on the Loading of Swept and Unswept Wings," NACA Rep. 1140, 1953.

[4]Diederich, F. W., and Foss, K. A., "Static Aeroelastic Phenomena of M, W, and Λ Wings," NACA RM L52J21, Feb. 1953.

[5]Gray, W. L., and Schenk, K. M., "A Method for Calculating the Subsonic Steady-State Loading on an Airplane with a Wing of Arbitrary Plan Form and Stiffness," NACA TN 3030, Dec. 1953.

14
Structural Design Considerations

The purpose of this chapter is to summarize various structural considerations that can have a significant impact on the design of the aircraft.

The selection of various design parameters as discussed in this chapter will affect the magnitude of structural design loads and the amount of structural weight that is built into the structure. The tendency at times is to use the "Be conservative" approach. This attitude may keep the flutter and loads engineers out of trouble but can lead to an overweight airplane that will not compete in the marketplace.

The subjects discussed in this chapter are gross weights, center of gravity limits, and selection of the positive and negative maximum normal force coefficients.

Proper consideration of each of these subjects will lead to an optimum design from a structures viewpoint, thus enhancing the performance of the airplane.

Although each of the subjects is discussed as an individual concern, the combined effect of gross weights, center of gravity positions, and design airspeeds must be kept in mind when proceeding with the analysis of the various structural components of the aircraft.

In addition, consideration will be given to the construction of V-n diagrams for maneuver and gust requirements.

14.1 Gross Weights

Two of the most significant factors contributing to the structural weight of an airplane are the selection of design gross weights and the center of gravity envelope.

The selection of design gross weights will, of course, be dependent on the missions chosen for the airplane and variables such as range requirements, takeoff and landing field lengths, and desired payload capability.

The gross weights that are usually considered for structural design purposes are *MTW*—maximum taxi gross weight (the maximum gross weight to which the aircraft can be loaded on the ground); *MTOW*—maximum takeoff gross weight (the maximum gross weight at the beginning of the takeoff run; this weight is designed for 6-ft/s sink rate, a 1.5-g maneuver at all landing flap positions, and a 2.0-g maneuver at all takeoff flap positions); *MLW*—maximum landing weight (the maximum landing weight at which the airplane is designed for 10-ft/s sink rate and a 2.0-g maneuver at all landing flap positions); *MZFW*—maximum zero fuel weight (the maximum weight at which the airplane may be loaded with zero usable fuel); *OEW*—operating empty weight (the weight of the airplane with zero payload and usable fuel); and OEW_{min}—minimum operating empty weight (the weight of the airplane with zero payload and usable fuel and with only minimum flight equipment on board, usually in a ferry configuration).

Some aircraft configurations, due to fuel tank configurations, may require special definitions such as the maximum in-flight weight for which outboard reserve fuel tanks may be empty or gross weights associated with fuel tanks located in the horizontal or vertical tails.

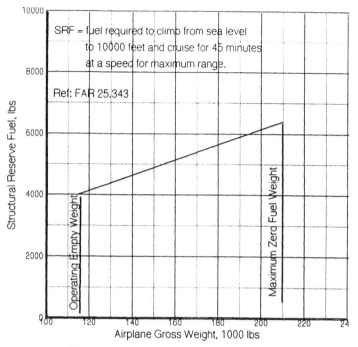

Fig. 14.1   Structural reserve fuel requirement.

14.1.1   Structural Reserve Fuel Conditions

Per the requirements of FAR 25.343, consideration must be given to structural reserve fuel conditions. These conditions are defined as the minimum fuel allowed when investigating conditions at full design load factors.

Structural reserve fuel conditions are bounded as follows:

$$(OEW + SRF) \text{ to } (MZFW + SRF)$$

where the structural reserve fuel SRF is defined by FAR 25.343(a) as the fuel required to climb from sea level to 10,000 ft and thereafter allowing 45 min of cruise at a speed for maximum range. A typical structural reserve fuel curve is shown in Fig. 14.1.

14.1.2   Maximum Design Payload

The maximum amount of payload required for design defines the maximum zero fuel weight of the airplane, hence,

$$\text{maximum design payload} = MZFW - OEW \qquad (14.1)$$

The actual design payload will be defined by the mission requirements of the airplane and is dependent on the number of passengers plus cargo that are selected as the design objectives. In terms of a pure cargo airplane, the maximum amount of cargo will be dependent on the design mission selected for the airplane. This is usually a compromise between the volume of cargo required vs the maximum design cargo weight desired.

14.1.3 Relationship Between Maximum Taxi and Landing Weights

In the selection of structural design gross weights, the relationship between maximum taxi and landing weights must be given special consideration.

Per the requirements of FAR 25.493, a factor of 1.2 is applied to the maximum landing weight in the calculation of the ground-handling braked-roll conditions as discussed in Sec. 7.4 of Chapter 7.

For most landing gear configurations, the braked-roll conditions are critical for design of the main gear oleos and supporting structure; therefore the selection of the maximum taxi weight may have a significant impact on the resulting structural weight of the landing gears (both main and nose gears).

For short-range aircraft where landing weight picks the design condition, the maximum taxi weight can be increased without affecting the landing gear structure designed by the braked-roll condition at maximum landing weight.

For aircraft designed for long-range operations where the maximum takeoff weight is of concern, the opposite is true, and increases in maximum landing weight may be obtained without affecting the landing gear structure designed by the braked-roll condition.

14.2 Center of Gravity Limits

The selection of the center of gravity limits for a given design are dependent on three concerns: 1) the loadability requirements for passengers and/or cargo, 2) the stability and control considerations and the effect on empennage size, and 3) the effect on structural loads and thus the sizing of structure and the resulting operating weight of the airplane.

The center of gravity limits for an airplane are usually defined in terms of the mean aerodynamic reference chord of the wing either as a ratio or in percent.

14.2.1 Center of Gravity Envelope Boundaries

The diagram shown in Fig. 14.2 is a composite of several airplanes with various limits being set by different concerns affecting not only the structure but also aerodynamic stability and elevator required for takeoff, landing, and dive recovery: 1) structural limits—limited by structural design considerations. 2) aerodynamic limits—forward center of gravity envelope usually limited by the amount of elevator available for takeoff rotation, landing, or dive recovery, and aft center of gravity limits usually based on stability considerations; 3) takeoff and landing limits—selected to meet the loading requirements for the airplane; 4) forward flight limit—selected for in-flight movement of passengers; and 5) thrust limits—nose gear steering limit with application of maximum takeoff thrust.

14.2.2 Selection of Center of Gravity Limits

The selection of the design center of gravity limits may be influenced by the structural load variation with gross weight and center of gravity position as shown in the constant load envelopes of Fig. 14.3.

The constant load envelopes calculated using simplified assumptions must be verified by a complete load analysis. Usually these analyses have proven that the simplified methods are acceptable for the selection of center of gravity envelopes.

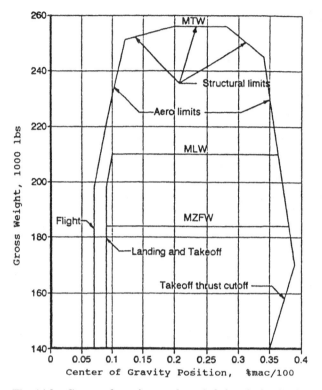

Fig. 14.2   Center of gravity envelope defining design limits.

14.2.2.1   Constant wing load vs center of gravity position.   The equation for constant wing load variation with center of gravity position may be derived from Eqs. (2.6) and (2.22):

$$\frac{\mathrm{d}L_w}{\mathrm{d}n_z} = W\left(\frac{x_t}{c_w} + 0.25 - CG\right)\Bigg/\left(\frac{x_t}{c_w} + \frac{\mathrm{d}C_M}{\mathrm{d}C_L}\right) \qquad (14.2)$$

where $\mathrm{d}L_w/\mathrm{d}n_z$ is the variation of wing load with load factor.

Assuming a constant wing load for a given flight condition, the relationship between gross weight and center of gravity position may be derived assuming constant Mach number, airspeed, and wing fuel:

$$W_i[x_t/c_w + 0.25 - CG_i] = \text{const} = k \qquad (14.3)$$

Constant wing load lines for a balanced maneuver condition are calculated in Table 14.1 using Eq. (14.3).

14.2.2.2   Constant balancing tail load vs center of gravity position.   The variation of balancing tail load with gross weight and center of gravity position

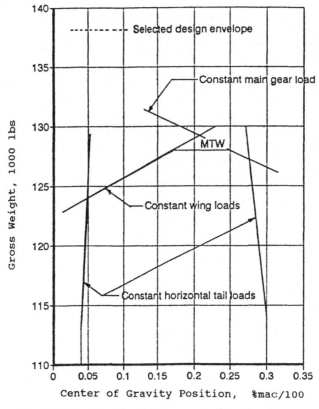

Fig. 14.3   Constant load envelope vs center of gravity.

can be derived from Eq. (2.22) assuming constant Mach number, airspeed, and wing fuel:

$$W_i \left[CG_i - 0.25 + \frac{dC_M}{dC_L} \right] = \text{const} = fk \qquad (14.4)$$

Constant horizontal tail load lines for a balanced maneuver condition are calculated in Table 14.2 using Eq. (14.4).

14.2.2.3 Constant nose gear load vs center of gravity position. The variation in nose gear ground loads with gross weight and center of gravity position can be determined from Eq. (7.13). In general, the nose gear will be critical

Table 14.1   Constant wing load lines for balanced
maneuver conditions calculated using Eq. (14.3)ᵃ

Condition	W_i, lb	CG_i, %c_w/100	k
1	128,000	0.170	519,552
2	125,830	0.100	519,552

ᵃFor this example $x_t/c_w = 3.979$.

Table 14.2   Constant horizontal tail load lines for balanced maneuver conditions calculated using Eq. (14.4)[a]

	Condition	W_i, lb	CG_i, $\%c_w/100$	fk
Forward center of gravity	1	113,000	0.0400	−10,227
condition	2	125,000	0.0487	−10,225
Aft center of gravity	1	114,000	0.293	18,525
condition	2	128,000	0.2752	18,522

[a]For this example $dC_M/dC_L = 0.1195$.

for forward center of gravity positions in the braked-roll condition. Assuming a constant vertical center of gravity position with gross weight, constant nose gear load lines may be calculated from Eq. (14.5):

$$W_i(B_i + E\mu) = \text{const} = k_{NG} \tag{14.5}$$

Constant nose gear load lines for a braked-roll condition are calculated in Table 14.3 using Eq. (14.5). Geometric parameters are defined in Fig. 7.1.

14.2.2.4   Constant main gear load vs center of gravity position.   Main gear constant load envelopes may be determined from two of the critical types of conditions that usually design main landing gears.

If the main gear is critical for the two-point braked-roll condition at landing weight, then the maximum taxi weight may be defined as shown in Eq. (14.6), which is derived from the requirement in FAR 25.493:

$$MTW = 1.2MLW \tag{14.6}$$

If the main gears are critical at *MTW*, i.e., *MTW* > 1.2*MLW*, then the upper limit will be braked-roll critical.

The variation of main gear load with airplane gross weight and center of gravity position for the ground turn condition may be derived from Eq. (7.53), assuming a constant vertical center of gravity position with gross weight:

$$W_i[A_i/2C + (BL_{cg} + 0.50E)/T] = \text{const} = k_{MG} \tag{14.7}$$

Constant main gear load lines for a ground turn condition are calculated in Table 14.4 using Eq. (14.7). Geometric parameters are defined in Fig. 7.1.

Table 14.3   Constant nose gear load lines for braked-roll conditions calculated using Eq. (14.5)[a]

Condition	W_i, lb	CG_i, $\%c_w/100$	B_i, in.	$k_{NG}\ 10^{-6}$
1	125,000	0.10	59.0	17.505
2	114,000	0	72.4	17.492

[a]For this example $E = 101.3$ in. and $\mu = 0.80$.

**Table 14.4   Constant main gear load lines for
ground turn conditions calculated using Eq. (14.7)[a]**

Condition	W_i, lb	CG_i, $\%c_w/100$	A_i, in.	k_{MG}
1	128,000	0.25	409.2	92,073
2	126,679	0.30	415.9	92,070

[a]For this example $E = 108.2$ in., $C = 448$ in., $T = 206$ in., and
$BL_{cg} = 0$.

14.3   Selection of Positive and Negative C_{Lmax}

The selection of positive and negative C_{Lmax} is important because of the effect
on structural design loads. Mach number and buffet limits at high-lift coefficients need to be considered in determining the flight boundaries required for
design.

The relationship between C_{Lmax} and C_{Nmax} will also be discussed.

14.3.1   Maximum Positive Lift Coefficient

In general, C_{Lmax} is determined from wind-tunnel tests, then corrected for full-
scale airplane effects using the experience obtained from flight testing similar
configurations. Analysis modification as necessary based on flight testing of the
actual airplane may be required.

The maximum lift coefficient for the airplane that can be developed is a function
of airplane center of gravity position CG. As noted in Eq. (14.8), C_{Lmax} varies
with the balancing tail load that can be developed:

$$C_{Lmax} = (C_{Lmax})_{WBN} + C_{Lt} \qquad (14.8)$$

where $(C_{Lmax})_{WBN}$ is the maximum lift coefficient tail-off, and C_{Lt} is the lift
coefficient due to horizontal tail.

The following relationship may be derived from Eq. (2.13):

$$C_{La}(CG - 0.25) + C_{M0.25} = BTLx_t/(q S_w c_w) \qquad (14.9)$$

The lift coefficient due to horizontal tail is defined by Eq. (14.10):

$$C_{Lt} = BTL/(q S_w) \qquad (14.10)$$

By inserting Eq. (14.10) into Eq. (14.9), one can obtain the relationship between
airplane lift coefficient and horizontal tail load coefficient:

$$C_{La}[CG - 0.25] + C_{M0.25} = C_{Lt}x_t/c_w \qquad (14.11)$$

The horizontal tail load coefficient may now be written as a function of airplane
center of gravity position:

$$C_{Lt} = [C_{La}(CG - 0.25) + C_{M0.25}]c_w/x_t \qquad (14.12)$$

The change in horizontal tail load coefficient due to change in airplane center of gravity position may now be derived:

$$\Delta C_{Lt} = C_{La}(CG - 0.25)c_w/x_t \qquad (14.13)$$

If Eq. (14.13) is substituted into Eq. (14.8), then the variation of maximum lift coefficient for the airplane can be obtained as a function of airplane center of gravity position:

$$C_{Lmax} = (C_{Lmax})_{0.25} + C_{Lmax}(CG - 0.25)c_w/x_t \qquad (14.14)$$

By collecting terms, one can calculate the maximum lift coefficient at any center of gravity position if the maximum lift coefficient at $0.25c_w$ is known from wind-tunnel or flight tests:

$$C_{Lmax} = (C_{Lmax})_{0.25}/[1 - (CG - 0.25)c_w/x_t] \qquad (14.15)$$

Thus for flight conditions with center of gravity positions forward of 0.25 mac the maximum lift coefficient is less than for center of gravity positions aft of 0.25 mac. This is illustrated in Table 14.5.

The significance of the example shown in Table 14.5 will become apparent in discussions on selecting critical conditions for the wing and horizontal tail.

14.3.2 Mach Number and Buffet Considerations

The variation of C_{Lmax} with Mach number may be determined from wind-tunnel tests. Tunnel limitations potentially make high angle-of-attack testing difficult or impossible because of the high forces placed on the model at dynamic pressures associated with high Mach numbers. This leads to the problem of determining C_{Lmax} at very high Mach numbers unless a variable density tunnel is used for these tests.

The variations of maximum lift coefficients with Mach number as obtained from wind-tunnel tests are corrected using experience from previous similar configurations, then verified by flight tests. The variation with Mach number of the maximum positive lift coefficient for a commercial jet transport is shown in Fig. 14.4. The buffet boundary limit at high Mach number has been verified by flight testing of the production airplane. The lift coefficients developed at Mach numbers associated with heavy buffet are not C_{Lmax} in the classic sense but rather the maximum lift coefficients that may be demonstrated in flight test.

Table 14.5   Variation of maximum lift
coefficient with airplane center of gravity
position, determined from Eq. (14.15), with
the horizontal tail incremental lift coefficient
obtained from Eq. (14.13)[a]

$CG, \%c_w/100$	C_{Lmax}	ΔC_{Lt}
0.10	1.154	−0.046
0.25	1.20	0
0.35	1.233	0.033

[a]For this example $c_w/x_t = 0.268$.

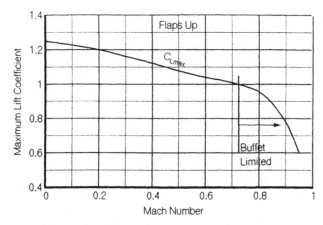

Fig. 14.4 Maximum lift coefficient variation with Mach number.

14.3.3 Relationship Between C_{Lmax} and C_{Nmax}

The relationship between C_{Lmax} and C_{Nmax} needs to be considered because the regulations, both FAR and JAR, define the positive left-hand boundaries of the V-n diagram in terms of C_{Nmax}:

$$n_z W = C_{Nmax} q S_w \qquad (14.16)$$

The relationship between C_{Lmax} and C_{Nmax}, shown in Eq. (14.17), may be determined by dividing Eq. (2.4) by $q S_w$:

$$C_{Nmax} = C_{Lmax} \cos \alpha_{wm} + C_{Dm} \sin \alpha_{wm} \qquad (14.17)$$

where C_{Dm} is the drag coefficient at C_{Lmax}, and α_{wm} is the wing reference angle of attack at C_{Lmax}.

The general practice is to assume for analytical purposes the equivalency shown in Eq. (14.18):

$$C_{Nmax} = C_{Lmax} \qquad (14.18)$$

The comparison of C_{Nmax} and C_{Lmax}, shown in Table 14.6 for two aircraft, indicates less than 1% difference between the two coefficients using the assumption of Eq. (14.18).

Table 14.6 Comparison of C_{Nmax} calculated from Eq. (14.17) and C_{Lmax}

Airplane	Mach no.	α_{wm}, deg	C_{Lmax}	C_{Dm}	C_{Nmax}	Ratio
A	0.40	11.5	1.24	0.165	1.248	1.006
A	0.78	6.0	0.94[a]	0.095	0.945	1.005
A	F40	20.0	3.40	0.640	3.423	1.007
B	0.45	11.0	1.09	0.076	1.084	0.995
B	0.80	7.0	0.84[a]	0.097	0.846	1.007

[a]Heavy buffet limited as confirmed by flight tests.

14.3.4 Determination of Airspeeds for Conditions at C_{Lmax}

The determination of airspeeds for conditions at C_{Nmax} requires solution of Eq. (14.16). If C_{Nmax} is a constant for a given airplane configuration, such as for flaps-down analyses, the dynamic pressure, hence velocity, can be readily obtained from Eq. (14.16):

$$q = n_z W / (C_{Nmax} S_w) \qquad (14.19)$$

If C_{Nmax} is nonlinear with Mach number, as is generally the case for flaps-up configurations, then the solution for dynamic pressure and velocity is more difficult.

Equations (14.20) and (14.21) define the relationships between dynamic pressure and airspeed and Mach number and airspeed:

$$q = \tfrac{1}{2}\rho V^2 \qquad (\text{lb/ft}^2) \qquad (14.20)$$

$$V = M c_s \qquad (\text{ft/s}) \qquad (14.21)$$

The speed of sound at any given altitude may be calculated from Eq. (14.22) as a function of pressure and density ratios[1]:

$$c_s = (\Gamma P_0 \delta / \rho_0 \sigma)^{\frac{1}{2}} \qquad (\text{ft/s}) \qquad (14.22)$$

Inserting Eqs. (14.21) and (14.22) into Eq. (14.20), one can derive the following relationship between dynamic pressure and Mach number:

$$q = \tfrac{1}{2}\rho M^2 (\Gamma P_0 \delta / \rho_0 \sigma) \qquad (14.23)$$

$$q = \tfrac{1}{2}\Gamma P_0 \delta M^2 \qquad (\text{lb/ft}^2) \qquad (14.24)$$

Using the definition of standard atmosphere from Ref. 1,

$$P_0 = 2116.216 \text{ lb/ft}^2 \qquad (14.25)$$

$$\Gamma = 1.4 \qquad (14.26)$$

one can write the equation for dynamic pressure as

$$q = 1481.35 M^2 \delta \qquad (\text{lb/ft}^2) \qquad (14.27)$$

The equation to determine the positive flight boundary can now be written in terms of the maximum lift coefficient, making the assumption that Eq. (14.18) is valid:

$$(n_z W / \delta)_{max} = 1481.35 M^2 S_w C_{Lmax} \qquad (14.28)$$

The variation of the parameter $(n_z W / \delta)_{max}$ may be plotted vs Mach number as shown in Fig. 14.5. This plot has the convenience of allowing easy determination of the Mach number for maximum lift coefficient conditions at a specified altitude, gross weight, and load factor.

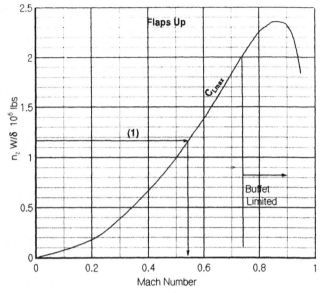

Fig. 14.5 Maximum lift capability, flaps up. The following procedure may be used to solve for the Mach number and airspeed at $C_{L\max}$ for a given altitude: 1) select $n_z W/\delta$ desired for the condition load factor, gross weight, and altitude; 2) determine Mach number from the curve as shown in the figure; 3) $V_e = 661.287 M \delta^{1/2}$, keas.

Using the procedure shown in Fig. 14.5, one can determine the Mach number for the condition from the graph, and airspeed may now be calculated as shown in Table 14.7.

Using the definition of equivalent airspeed as discussed in Chapter 16, shown by Eq. (16.6), and using Eqs. (14.20) and (14.21), equivalent airspeed may be determined as a function of Mach number using Eq. (14.29):

$$V_e = (\Gamma P_0/\rho_0)^{\frac{1}{2}} M \delta^{\frac{1}{2}} = c_{so} M \delta^{\frac{1}{2}} \qquad (14.29)$$

Table 14.7 Determination of airspeeds for conditions at $C_{L\max}$ using Eq. (14.30) and the procedure shown in Fig. 14.5

Altitude, ft	δ	W, lb	n_z	$n_z W/\delta$,[a] lb 10^{-6}	Mach no. at $C_{N\max}$	V_e, keas
35,000	0.2353	240,000	2.309[b]	2.3556	0.860	275.9
30,000	0.2970	240,000	2.5	2.0202	0.740	266.7
20,000	0.4595	240,000	2.5	1.3057	0.580	260.0
0	1.0	240,000	2.5	0.600	0.378	250.0

[a]See Eq. (14.29) where $S_w = 2500$ ft^2 for this example.
[b]Maximum load factor that can be developed at this altitude for the gross weight shown.

The term defined within the parentheses is the speed of sound at sea level for a standard day. Using the value defined in Ref. 1 and converting to knots, one can now calculate equivalent airspeed at any altitude using Eq. (14.30):

$$V_e = 661.287 M \delta^{\frac{1}{2}} \qquad \text{(keas)} \qquad (14.30)$$

The speed of sound at sea level is $c_{so} = 1116.89$ ft/s (Ref. 1), knots $= 0.868382$ (miles/h) (Ref. 2), and $c_{so} = (60/88)(0.868382) \, 1116.89 = 661.287$ keas.

14.3.5 Maximum Negative Lift Coefficient

Wind-tunnel testing to determine negative C_{Lmax} may become difficult due to the model mounting system used in the tunnel and the ability to pitch the model into significant negative angles of attack. The common practice is to assume that the maximum negative lift coefficient is a factor of the positive C_{Lmax}, as defined in Eq. (14.31):

$$\text{negative } C_{Lmax} = -k_L (\text{positive } C_{Lmax}) \qquad (14.31)$$

The factor k_L varies within the industry from 0.6 to 1.0. The rationale for the 1.0 factor is that the airplane has the same maneuver capability in the negative regime as it does in the positive regime. This is conservative but is not realistic in that the main lifting surface (namely the wing) should have a greater lifting capability in the positive direction for which the wing is designed.

14.4 V-n Diagrams

The criteria in FAR 25.333 state that the strength requirements must be met at each combination of airspeed and load factor on and within the boundaries of the representative maneuvering and gust envelopes.

Consideration will be given in this chapter to flight maneuvering and gust envelopes as required by the civilian regulatory agencies.

14.5 Maneuvering Envelope

The maneuver envelope, commonly called a V-n diagram, is described as follows:

1) The left-hand boundary depends on the maximum positive and negative static normal force characteristics of the airplane.

2) The right-hand boundary depends on the airplane design dive or flap placard airspeeds.

3) The upper and lower boundaries are defined by the design maneuver load factor requirements for the airplane.

4) At altitudes where the airplane may be limited by buffet considerations, the positive boundary may be defined by the lift coefficient at which the maneuvering capability is limited by heavy buffet.

These envelopes must be used in determining the airplane structural operating limitations as specified in FAR 25.1501.

Table 14.8 Maneuvering envelope calculations for flaps-down V-n diagram[a,b]

Flap configuration	W, lb	C_{Nmax}	n_z	V_e, keas
1	300,000	1.64	0.5	103.9
Takeoff			0.75	127.2
			1.0	146.9
			1.5	179.9
			2.0	207.8
30	250,000	2.53	0.5	76.4
Landing			0.75	93.5
			1.0	108.0
			1.5	132.3
			2.0	152.7
30	300,000	2.52	0.5	83.8
Landing			0.75	102.6
			1.0	118.5
			1.5	145.2

[a]The relationship between load factor and airplane maximum normal force coefficient is based on Eq. (14.16) modified as shown here: $n_z W = C_{Nmax} V_e^2 S_w / 295$ [Eq. (14.16a)].
[b]The examples shown in this table are based on an assumed area and normal force coefficients; $S_w = 2500 \text{ ft}^2$.

14.5.1 Maneuvering Envelopes Flaps Up and Down

At altitudes where the airplane has the lift capability to obtain design maneuver load factors, the upper left boundary may be calculated using Eq. (14.19) if the maximum static normal force coefficient does not vary with Mach number. Calculations for a flaps-down V-n diagram where the maximum normal force coefficient is constant for a given flap position are shown in Table 14.8.

Maneuvering envelopes are shown in Fig. 14.6 for flaps-down conditions and Fig. 14.7 for a flaps-up condition at an altitude where the design maneuver load factors can be attained.

For conditions where the maximum normal force coefficient is a function of Mach number, the upper left boundary may be calculated using the procedure discussed in Sec. 14.3.4. Calculations for a flaps-up V-n diagram where the maximum normal force coefficient is a function of Mach number are shown in Table 14.9. The maneuver capability at altitude calculated using this procedure is shown in Fig. 14.8.

14.5.2 Maneuvering Envelopes for High Altitudes

At high Mach numbers the airplane may be limited by heavy buffet, as discussed in Sec. 14.3.2.

The high-altitude maneuver envelope shown in Fig. 14.8 is based on the calculations given in Table 14.9.

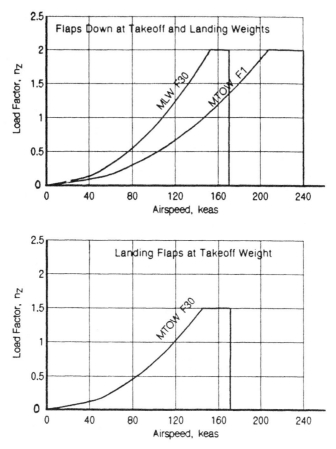

Fig. 14.6   Maneuvering envelopes with flaps down.

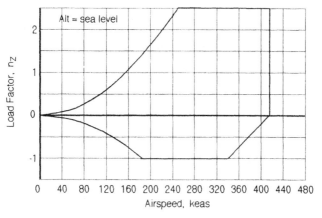

Fig. 14.7   Maneuvering envelope with flaps up at low altitude.

Table 14.9   Maneuvering envelope calculations for flaps-up V-n diagram where the relationship between load factor and airplane maximum normal force coefficient is shown in Eq. (14.28), and equivalent airspeeds are calculated using Eq. (14.30)[a]

Altitude, ft	δ	W, lb	n_z	$n_z W/\delta$, lb 10^{-6}	Mach[b] no. at $C_{N\max}$	V_e, keas
35,000	0.2353	240,000	1.0	1.020	0.505	162.0
			1.5	1.530	0.634	203.4
			2.0	2.040	0.746	239.3
			2.309	2.3556	0.860	275.9
0	1.0	240,000	1.0	0.240	0.234	154.7
			1.5	0.360	0.288	190.5
			2.0	0.480	0.335	221.5
			2.5	0.600	0.378	250.0

[a]The examples shown in this table are based on an assumed area and normal force coefficients. The normal force coefficients are shown as a function of Mach number in Fig. 14.4; $S_w = 2500$ ft^2.
[b]Obtained from Fig. 14.5 at $n_z W/\delta$.

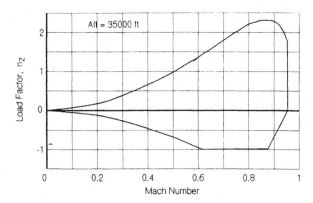

Fig. 14.8   Maneuvering envelope with flaps up at high altitude.

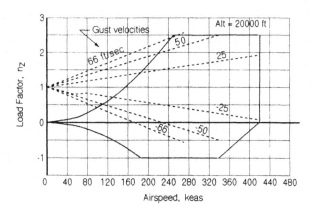

Fig. 14.9   Gust envelope, flaps up at 20,000 ft; gust velocities are defined in Table 5.2.

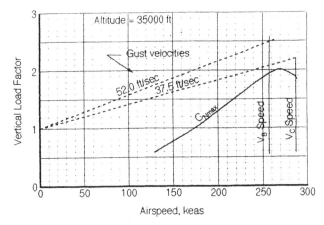

Fig. 14.10 Gust envelope, flaps down at 35,000 ft; gust velocities as defined in Table 5.2.

14.6 Gust Envelope

The gust envelope, commonly called a V-g diagram, is determined in a similar manner as the maneuvering envelope, except the boundaries are determined by the gust load factors calculated at V_B, V_C, and V_D speeds.

During the FAR/JAR harmonization process it was proposed that FAR/JAR 25.341 be rewritten, eliminating the gust formula for calculating load factors for discrete gust conditions. This proposal deletes the requirement for the gust envelope as previously specified in FAR 25.333(c).

14.6.1 Historical Approach for Gust Load Envelopes

The traditional approach for calculating vertical load factors using the revised gust formula is discussed in Sec. 5.2. The resulting load factors are usually superimposed on the maneuver envelope, calculated as discussed in Sec. 14.5.

Gust envelopes are shown in Figs. 14.9 and 14.10 for two altitudes, where load factors are calculated using Eq. (5.4). The design gust velocities are based on the criteria of FAR 25.241(c) before harmonization as summarized in Table 5.2. For simplification purposes, the V-g diagrams are drawn assuming linear variation of load factor with airspeed. The gust load factors calculated at the design speeds V_B, V_C, and V_D include Mach number and static aeroelastic effects.

The gust envelope in Fig. 14.10 for high altitude shows load factors that exceed the maneuvering capability of the airplane. This phenomenon may not actually occur in real flight as the aerodynamic characteristics become nonlinear in the flight regime significantly above the initial buffet lift capability not shown in these figures. For design purposes the gust load factors are critical at the lower altitude condition.

References

[1]Anon., "Manual of the ICAO Standard Atmosphere, Calculations by the NACA," NACA TN 3182, May 1954.

[2]Eshbach, O. W., *Handbook of Engineering Fundamentals,* 1st ed., Wiley, New York, 1945.

15
Structural Design Airspeeds

The selection of structural design airspeeds not only is a regulatory concern but also has a major impact on the resulting design loads, hence on the structural weight, of the airplane.

15.1 Cruise and Dive Speeds

The structural design cruise airspeed of commercial aircraft, V_C, is selected to envelop the maximum desired operational speeds, flaps up, which will become the maximum speed in which the airplane can be operated in commercial revenue service. Flight testing will be allowed to the maximum design airspeed, which is called the dive speed V_D.

The Airplane Flight Manual (AFM) has a section containing the operating limits of the airplane per the requirements of FAR 25.1583, as follows: "The maximum operating limit speed V_{MO}/M_{MO} and a statement that this speed limit may not be deliberately exceeded in any regime of flight (climb, cruise, or descent) unless a higher speed is authorized for flight test or pilot training."[5]

15.1.1 Historical Perspective

Before December 1964, two other airspeeds were incorporated in the AFM; these speeds were V_{NO} and V_{NE}. Definitions of these airspeeds were carried over from CAR 4b.711 and 4b.712[1] as follows:

> V_{NO} Speeds: The normal operating limit speed V_{NO} shall be established not to exceed the design cruising speed V_C chosen in accordance with 4b.210(b)(4) and sufficiently below the never-exceed speed V_{NE} to make it unlikely that V_{NE} would be exceeded in a moderate upset occurring at V_{NO}.
> V_{NE} Speeds: To allow for possible variations in the airplane characteristics and to minimize the possibility of inadvertently exceeding safe speeds, the never-exceed speed V_{NE} shall be established sufficiently below the lesser of ... [see CAR 4b.711(1) and (2)].

NASA VGH (velocity, load factor, and altitude) recorders were installed on 12 types of turboprop and turbojet aircraft during the period 1960–61. Analysis of these data showed that the operation speeds V_{NO} were being exceeded significantly more frequently than had been experienced in operations of piston-engine transports.[2-4]

Because of the structural implications of these studies, the regulatory agencies and industry dropped the use of V_{NE} for commercial transports certified under FAR 25.[5] In 1964 the maximum operating limit speed V_{MO} was introduced in FAR 25.1505; i.e., "V_{MO} Speeds: The maximum operating limit speed (V_{MO}/M_{MO} airspeed or Mach number, whichever is critical at a particular altitude) is a speed that may not be deliberately exceeded in any regime of flight (climb, cruise, or descent), unless a higher speed is authorized for flight test or pilot training."[5]

The "red line" on the airspeed indicator that was formerly associated with V_{NE} is now associated with V_{MO}. Modern technology in airspeed indicating systems has allowed V_{MO}, hence the "red line," to vary with altitude, but in all cases it does not exceed the structural design V_C speed.

15.1.2 Structural Design V_C Speed

The selection of the structural design V_C speed must comply with the minimum standards set for it in FAR 25.335(a); i.e., 1) the minimum value of V_C must be sufficiently greater than V_B to provide for inadvertent speed increases likely to occur as a result of severe atmospheric turbulence. 2) In the absence of a rational investigation substantiating the use of other values, V_C may not be less than $V_B + 43$ keas. However, it need not exceed the maximum speed in level flight at maximum continuous power for the corresponding altitude. 3) At altitudes where V_D is limited by Mach number, V_C may be limited to a selected Mach number.

In the selection of V_C speeds for commercial transports the application of FAR 25.335(a)(2) has applied only to the lower altitudes where V_C is defined in terms of knots calibrated airspeed or knots equivalent airspeed. At higher altitudes where V_C is limited to a selected Mach number, as allowed in FAR 25.335(a)(3), the 43-keas margin is not maintained.

The calculation of the airspeed margins between V_B and V_C are shown in Table 15.1.

From a historical perspective, the rationale for the 43-keas speed margin (other than being 50 mph) is unknown to the author. As noted in Table 15.1, the application of a head-on gust at V_B speed would not exceed V_C at altitudes where V_C is not Mach limited. This would then meet the intent of FAR 25.335(a)(1).

15.1.3 Structural Design V_D Speed

The selection of the structural design V_D speed must comply with the minimum standards set forth in FAR 25.335(b), whereby V_D must be selected so that V_C/M_C is not greater than $0.8\ V_D/M_D$, or the minimum speed margin between V_C/M_C and V_D/M_D is the greater of the following values:

1) From an initial condition of stabilized flight at V_C/M_C, the airplane is upset, flown for 20 s along a flight path 7.5 deg below initial path, and then pulled up at a load factor of 1.5 g. The speed increase occurring in this maneuver may be calculated if reliable or conservative aerodynamic data are used. Power as specified in FAR 25.175(b)(1)(iv) is assumed until the pull-up is initiated, at which time power reduction and the use of pilot-controlled drag devices may be assumed.

2) The minimum speed margin must be enough to provide for atmospheric variations (such as horizontal gusts and penetration of jet streams and cold fronts) and for instrument errors and airframe production variations. These factors may be considered on a probability basis. However, the margin at altitude where M_C is limited by compressibility effects may not be less than $0.05M$.

In the practical application of FAR 25.335(b)(1), it was the intent of this regulation that the speed margin due to an upset in pitch was to be determined by analysis only and would not be demonstrated by flight tests. The requirement for flight testing as given in Ref. 2 considers the operational upsets expected to occur in service to cover pitch, roll, yaw, and combined upsets. Successful completion of these flight tests will more than adequately substantiate the margin between V_C and V_D speeds used for structural design of the airplane.

Table 15.1 Airspeed margins between V_B and V_C where the airplane's speeds are defined in terms of knots calibrated airspeed

Altitude, ft	V_C		V_B		$\blacktriangle V$,[a]	U_{de},[b]
	kcas	keas	kcas	keas	keas	keas
0	350	350	290	290	60	39.1
20,000	350	338.1	290	282.9	55.2	39.1
25,000	350	333.2	290	279.7	53.5	36.3
30,000	$M0.86$	310.0	290	276.1	33.9[c]	33.6
35,000	$M0.86$	275.9	$M0.78$	250.3	25.6[c]	30.8
40,000	$M0.86$	244.7	$M0.78$	222.0	22.7[c]	28.0

[a]$\blacktriangle V = V_C - V_B$.
[b]66 ft/s eas up to an altitude of 20,000 ft. See FAR 25.341(a)(1).
[c]At altitudes where V_C speeds are Mach limited, the certifying agencies have allowed a deviation from the requirements of FAR 25.335(a)(2).

15.1.4 Upset Analysis for Determining V_D Speed

The equation for determining the change in airspeed during prolonged descents is obtained from Ref. 6:

$$\frac{dV}{dt} = g[\sin(\tau_{\text{req}} + \tau_0) + T/W - C_D/C_L] \tag{15.1}$$

where V is the true airspeed (ft/s), t is the time (s), $\tau_{\text{req}} = 7.5$ deg per FAR 25.335(b)(1), τ_0 is the stabilized angle for flight at the airspeed from which the upset is started ($t = 0$), $\tau_0 = 0$ for stabilized flight when the engine thrust required for flight is equal or less than thrust available, g is the acceleration of gravity (32.2 ft/s²), T is the engine thrust (lb), W is the airplane gross weight (lb), C_D is the airplane drag coefficient, and C_L is the airplane lift coefficient.

Equation (15.1) is based on a constant dive angle from the stabilized condition from which the upset is performed. If the assumptions are made that engine thrust and gross weight are the values at the beginning of the upset, the only variables in the problem are the drag and lift coefficients C_D and C_L.

The lift coefficient may be calculated from Eq. (15.3):

$$n_z W = C_L q S_w \tag{15.2}$$

$$C_L = n_z W/(q S_w) \tag{15.3}$$

Since the dynamic pressure q varies with time during the upset, it follows that C_L will vary with time. The drag coefficient C_D is a function of airplane lift coefficient and Mach number, and so it will be time dependent in this analysis:

$$C_D = f(C_L, \text{Mach number}) \tag{15.4}$$

Per FAR 25.335(b)(1), the load factor n_z would be essentially equal to 1 g for the first 20 s after the initiation of the upset. After 20 s the load factor would vary during the pull-out up to a maximum of 1.5 g, at which time the engine

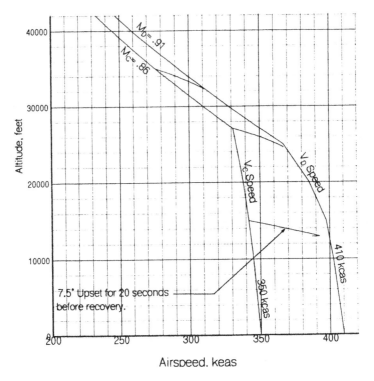

Fig. 15.1   Upset speeds as defined by FAR 25.335(b)(1).

thrust T may be reduced and speedbrakes, if available, may be deployed (hence contributing to the drag coefficient C_D).

The solution of Eq. (15.1) can be readily accomplished using high-speed computers, having the variation of drag with lift coefficient and Mach number. The upset speeds shown in Fig. 15.1 are based on the analysis shown in Table 15.2.

15.2   Maneuvering Speeds

The purpose of this section is to provide historical background on the determination of the airplane maneuvering speed V_A as set forth in FAR 25.335(c).

15.2.1   Maneuvering Speed Definition

The maneuvering speed V_A is the maximum airspeed at which the pilot may apply a single application of any one control surface to its maximum angle, limited by 300-lb of pilot effort, stops, or blowdown. This control surface motion is intended to be in one direction only and does not include oscillatory inputs.

15.2.2   Historical Perspective

In the era before the introduction of commercial jet transports in the United States in 1956, the airplane maneuvering speed was defined as the airspeed at the upper left-hand corner of the V-n diagram in the flaps-up configuration. In essence, this was the lowest speed at which the maximum design positive load

Table 15.2 Airplane upset analysis used to substantiate V_D speeds calculated using Eq. (15.1); the dive angle is 7.5 deg from the initial stabilized angle for flight at the beginning of the upset

	Time, s	Altitude, ft	Airspeed, keas	Mach	Thrust, lb/engine	Speedbrakes
Altitude, ft = 35,000	5.0	34,346	285.9	0.877	8,761	0
V_e, keas = 275.9	10.0	33,678	295.0	0.891	8,906	0
Mach = 0.86	15.0	32,999	303.3	0.902	9,052	0
Initial dive angle,	20.0	32,310	310.8	0.909	9,201	0
deg = 1.38	20.5	32,245	311.0	0.908	0	Up
Upset angle, deg = 7.5	21.0	32,188	308.3	0.900	0	Up
Gross weight,	21.5	32,140	305.7	0.891	0	Up
lb = 215,000	22.0	32,100	303.1	0.882	0	Up
Thrust, lb/engine	23.0	32,037	297.8	0.866	0	Up
= 8620	24.0	31,991	292.7	0.850	0	Up
Altitude, ft = 30,000	5.0	29,369	320.2	0.876	9,956	0
V_e, keas = 310	10.0	28,726	329.8	0.889	10,217	0
Mach = 0.86	15.0	28,072	338.3	0.899	10,482	0
Initial dive angle,	20.0	27,411	345.8	0.905	10,750	0
deg = 0.89	20.5	27,348	345.8	0.904	0	Up
Upset angle, deg = 7.5	21.0	27,294	342.2	0.893	0	Up
Gross weight,	21.5	27,249	338.7	0.883	0	Up
lb = 216,000	22.0	27,212	335.2	0.873	0	Up
Thrust,	23.0	27,154	328.3	0.854	0	Up
lb/engine = 9700	24.0	27,115	321.7	0.836	0	Up

factor could be attained as limited by the maximum lift capability of the airplane, flaps up:

$$V_A = (n_z W 295 / C_{N\max} S_w)^{\frac{1}{2}} \qquad \text{(keas)} \qquad (15.5)$$

where n_z is the maximum design positive maneuver load factor, W is the maximum design flight gross weight (lb), $C_{N\max}$ is the maximum design normal force coefficient, and S_w is the wing reference area (ft²).

For commercial airplanes designed before 1956 the effect of compressibility on the maximum lift coefficient was ignored, and hence the maneuvering speed could be defined in terms of the 1-g stall speed:

$$V_A = V_{s1g} \sqrt{(n_z)} \qquad \text{(keas)} \qquad (15.6)$$

where V_{s1g} is the airplane stall speed at $n_z = 1.0$ (keas).

15.2.3 Present Practice

Since the advent of modern commercial jets in 1956, with $C_{N\max}$ varying with Mach number, the method of determining V_A speeds had to be changed. At high altitudes where it became unrealistic to pull the maximum design load factor, the definition of the maneuvering speed became impossible to determine in the traditional sense. There is no upper left-hand corner of the V-n diagram as shown in Fig. 15.2. The reason for the dropoff in load factor at high Mach numbers is due to maximum lift capability being limited by heavy buffeting.

Load factors are calculated in Table 15.3 for incompressible and compressible $C_{N\text{max}}$ at sea level. The V_A speed calculated using Eq. (15.6), as shown in Fig. 15.3, is in agreement with the upper left-hand corner of the V-n diagram based on incompressible $C_{N\text{max}}$.

To maintain the philosophy and need for having a maneuvering speed at these altitudes, it was decided by the FAA and industry representatives to determine V_A speeds defined by Eq. (15.6). These airspeeds are shown in Figs. 15.2 and 15.3.

Since V_A speeds vary with gross weight, the general practice is to select speeds as a function of altitude, including the anticipated growth in gross weight for the airplane. This procedure allows for calculating loads only once on surfaces like the vertical tail for which the loads normally do not vary with airplane gross weight. This does not apply for a "T" tail configuration.

15.3 Flap Placard Speeds and Altitude Limitations

Flap placard speeds are some of the most significant airspeeds that affect the design of commercial aircraft. The higher the placards, the more structure is required in the trailing-edge flaps and attachment structure. Selection of these speeds must be done carefully with an understanding of regulatory minimum requirements vs the basic performance goals for the airplane.

15.3.1 Historical Perspective

In the early days before 1953, U.S. aircraft manufacturers were required to design to Ref. 1. These airplanes essentially had two flap positions: up and landing positions. It was required to maintain adequate speed margins such that stall would not occur when retracting flaps at placard speed. The criteria of CAR 4b were defined by Eqs. (15.7) and (15.8):

Flaps retracted at design landing weight:

$$V_F = 1.4V_s \tag{15.7}$$

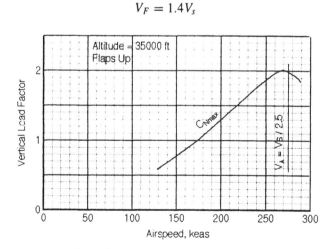

Fig. 15.2 V_A speed at high altitude.

Table 15.3 Maneuver envelope calculated using incompressible and compressible $C_{N\max}$ where the compressible analysis is the same as shown in Table 14.9; $S_w = 2500$ ft²

	Altitude, ft	δ	W, lb	n_z	$C_{N\max}$	V_e, keas
Compressible analysis	0	1.0	240,000	1.0	1.183	154.7
	——	——	——	1.5	1.171	190.5
	——	——	——	2.0	1.154	221.5
	——	——	——	2.5	1.133	250.0
Incompressible analysis[a]	0	1.0	240,000	1.0	1.200	153.6
	——	——	——	1.5	1.200	188.1
	——	——	——	2.0	1.200	217.3
	——	——	——	2.5	1.200	242.9

[a] Incompressible analysis based on $M=0.20$ data as obtained from a low-speed wind tunnel.

Landing flaps at design landing weight:

$$V_F = 1.8V_s \qquad (15.8)$$

whichever is the greater.

Calculations of flap placard speeds using Eqs. (15.7) and (15.8) are shown in Table 15.4.

15.3.2 Flap Placard Speed Requirements

In 1961 Ref. 1 was amended to delete the flaps-up requirement and replace it with a takeoff flap requirement. The landing flap requirement defined by Eq. (15.8) was maintained in the regulations:

Takeoff flaps at maximum takeoff weight:

$$V_F = 1.6V_s \qquad (15.9)$$

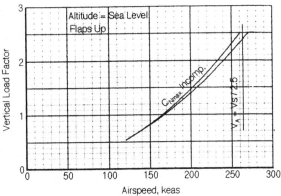

Fig. 15.3 V_A speed at sea level.

Table 15.4 Flap placard speed calculation based on Boeing 377
with four piston engines designed per CAR 4b, effective Nov. 9,
1945; $MTOW$ = 147,000 lb and MLW = 130,000 lb

	Gross weight, lb	Flap position, deg	V_s, mph	$1.4V_s$, mph	$1.8V_s$, mph
Landing flaps	130,000	0	133	186	——
	130,000	45	106	——	191[a]
Takeoff flaps	147,000	0	141	197[b]	——
	147,000	25	123	——	221[a]

[a] The placard speeds for the trailing-edge flaps would be chosen as indicated.
[b] Suppose the takeoff flap condition zero flap speeds were 230 mph instead of the 197 shown. The airplane would not have adequate stall margin when the flaps were retracted after takeoff if the takeoff flap placard is 221 mph.

where V_s is the stall speed at the flap position and design weight under consideration (these stall speeds may occur at load factors less than 1 g depending on how the flight test was flown), and V_F is the flap placard speed.

15.3.3 FAR 25 and Multiposition Flaps

With the introduction of multiposition flaps like those on the 707 and 727 aircraft, the flaps-up rule defined in Eq. (15.7) was eliminated in FAR 25 and replaced with the requirement whereby the transition from one flap position to another must provide adequate operational speed margins during the flap transition [see FAR 25.335(e)(1)].

Per FAR 25.335(e)(3), minimum flap placard speeds are now defined using Eqs. (15.8) and (15.9).

15.3.4 Flap Placard Considerations for CAA Certification

The determination of flap placard speeds for certification by the British Civil Aviation Authority (CAA) is done in a similar manner as that for FAA certification, except the following must also be considered:

$$V_F > 1.4V_s \qquad (15.10)$$

where V_s is the stall speed for the next gated flap position (a gate being a physical stop at a flap detent). The pilot must take physical action to move the handle beyond this position, much like changing gears in a car.

The purpose of this rule was to ensure adequate speed margins at the gated positions in the event that the pilot selected an incorrect lower flap setting during the retract sequence.

15.3.5 En Route Flaps

FAR 25.345(c) requires consideration of use of flaps en route where flaps-up design load factors, $n_z = 2.5$ (for gross weights greater than 50,000 lb), must be used for design, flaps down. This became a concern of the regulatory agencies in the operation of the early jet transports as some airlines slowed the aircraft to flap placard speeds for minimum flaps even at altitudes as high as 35,000 ft before starting descent. The FAA was concerned that compressibility effects, which were neglected for sea level conditions, might be significant at these altitudes even though flap placard speeds were stated in calibrated or indicated airspeeds.

Normally aircraft flap placard speeds are quoted in equivalent airspeed for the determination of loads but are used as calibrated or indicated airspeeds for the placards in the AFM. This provides some conservatism at altitude but does not adequately cover compressibility effects.

The addition of the 20,000-ft altitude limitation reinforced the argument with the FAA that flaps should not be used for en route operation of the airplane.

15.3.6 Flap Load Alleviation

Depending on the type of flap system chosen for design, consideration may be given to reducing the loads on trailing-edge flaps by providing a flap retractor that would move the flaps from a more extended position to a lesser position. This may be accomplished using speed sensing devices, thus allowing for lesser design flap placard speeds as allowed by FAR 25.335(e)(2). Margins must be provided for operational considerations such as approach speed requirements plus turbulence.

A typical operational speed may be defined as

$$V_{\text{operational}} = V_{\text{ref}} + 20 \text{ kn} \tag{15.11}$$

where

$$V_{\text{ref}} = 1.3V_s \tag{15.12}$$

The minimum speed for design of a flap retractor should exceed the airspeed defined in Eq. (15.11) so that the flap placard speed would not be exceeded during a full power go-around in an aborted landing approach.

The use of a multistage flap retractor system would allow the flaps to retract from the full down position to a lesser position such that the impact on design loads would be minimized by the problem of using 1-g stall speeds in defining placard speed requirements.

A flap system driven by hydraulic actuators may be designed at lower placard speeds because the flaps will blow back at the selected design speed, thus alleviating loads to a given design level.

15.3.7 Altitude Limitations

In the early days of the 707 airplane, operations were allowed in the flaps-down mode at all altitudes up to maximum certificated altitude (42,000 ft in the case of the 707 airplane). Flap loads were calculated on the basis of incompressible data at the equivalent airspeed for the flap position under consideration. The airplane placard was established in terms of knots calibrated or indicated airspeed.

When the design of the 727 airplane was under way, the FAA raised the question of compressibility effects on flap loads (including leading-edge Krueger flaps and slats).

An altitude limit of 20,000 ft was accepted as a reasonable limitation to remove the need for consideration of compressibility effects on the high-lift device loads. The 707, 727, 737, 747, and 757 flap loads were substantiated during instrumented flight testing up to flap placard speeds in the altitude range of 10,000–15,000 ft.

Compressibility correction factors may be computed from Eq. (15.13)[7]:

$$F_c = 1/[1 - M^2 \cos^2(SA)]^{\frac{1}{2}} \qquad (15.13)$$

where SA is the trailing-edge flap swept-back angle.

The effect of compressibility on flap loads for two airplanes is shown in Table 15.5.

15.4 Gust Design Speeds

The criteria for gust design speeds have evolved since the advent of commercial jet transports in 1953. The change in calculating these speeds came about due to operational experience during the early years of the first generation of commercial jets.

There are three speeds that involve the determination of design loads due to gusts: V_B, V_C, and V_D. The latter two speeds are the maximum design cruise speed and the design dive speed discussed in Secs. 15.1.2 and 15.1.3.

Table 15.5   Effect of compressibility on trailing-edge flap loads[a]

Flap position	Altitude, ft	V_F, kcas	V_F, keas	Mach	F_c[b]	$(q F_c)$, lb/in.^{2}	Ratio[c]
Airplane A inboard trailing-edge flaps (SA = 10.61 deg)							
5	0	240	240.0	0.363	1.070	1.451	1.000
	20,000	240	235.7	0.526	1.168	1.527	1.053
	35,000	240	229.1	0.714	1.404	1.735	1.196
25	0	190	190.0	0.287	1.042	0.886	1.000
	20,000	190	187.8	0.419	1.097	0.912	1.029
	35,000	190	184.3	0.574	1.211	0.968	1.093
40	0	175	175.0	0.265	1.036	0.747	1.000
	20,000	175	173.3	0.386	1.081	0.764	1.023
	35,000	175	170.5	0.531	1.172	0.802	1.074
Airplane B inboard trailing-edge flaps (SA = 0 deg)							
1	0	240	240.0	0.363	1.073	1.455	1.000
	20,000	240	235.7	0.526	1.176	1.538	1.057
	35,000	240	229.1	0.714	1.428	1.764	1.213
20	0	202	202.0	0.305	1.050	1.009	1.000
	20,000	202	199.4	0.445	1.117	1.045	1.036
	35,000	202	195.2	0.608	1.260	1.130	1.120
30	0	175	175.0	0.265	1.037	0.748	1.000
	20,000	175	173.3	0.386	1.084	0.766	1.025
	35,000	175	170.5	0.531	1.180	0.808	1.080

[a]Both airplanes have a 20,000-ft-altitude flaps-down placard in the AFM.
[b]Compressibility factors are calculated using Eq. (15.13).
[c]Ratio = $(q F_c)_{alt}/(q F_c)_{sea\ level}$.

15.4.1 Definition of V_B Speed

The V_B speed is defined as the design speed for maximum gust intensity. Per FAR 25.335(d)[5] the following apply.

1) The V_B speed may not be less than the speed determined by the intersection of the line representing the maximum lift $C_{N\max}$ and the line representing the rough air gust velocity on the gust V-n diagram, or $V_{s1}\sqrt{n_g}$, whichever is less, where a) n_g is the positive airplane gust load factor due to gust, at V_C (in accordance with 25.341), and at the particular weight under consideration; and b) V_{s1} is the stalling speed with the flaps retracted at the particular weight under consideration.

2) V_B speed need not be greater than V_C.

The definition in JAR 25.335(d) is essentially the same as is stated for FAR 25.335(d).

In the prejet days V_B speed was always defined by the intersection of the V_B gust line with the $C_{N\max}$ line on the V-n diagram. Compressibility effects were neglected, and the upper left side of the V-n diagram could be defined as shown in Eq. (15.14):

$$V_{@n} = V_{s1g}\sqrt{n_z} \qquad (15.14)$$

where

$$V_{s1g} = [295W/C_{N\max}S_w]^{\frac{1}{2}} \qquad \text{(keas)} \qquad (15.15)$$

n_z is the load factor, S_w is the wing area (ft^2), W is the gross weight (lb), and $C_{N\max}$ is the maximum normal force coefficient for the condition (constant with load factor).

An example of the determination of V_B using a constant $C_{N\max}$ is shown in Fig. 15.4, along with a $C_{N\max}$ that varies with Mach number. As noted, V_B would increase over the incompressible solution by 14 keas. In this example the assumption is made that the lift curve slope used in calculating gust load factors does not vary with Mach number.

With the coming of high-speed jet transports, which have a significant variation of $C_{N\max}$ with Mach number, V_B speeds could not be determined at high altitudes because the V_B and $C_{N\max}$ lines on the V-n diagram did not intersect; hence an

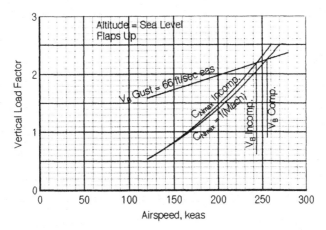

Fig. 15.4 V_B speed using incompressible $C_{N\max}$.

empirical criterion was established to allow determination of V_B speeds at high altitudes. The equation that was developed is similar to Eq. (15.14), modified as shown in Eq. (15.16):

$$V_{Bmin} = V_{s1}\sqrt{n_g} \qquad (15.16)$$

where V_{s1} is the stalling speed as defined by FAR 25.335(d)(1)(ii), and n_g is the positive gust load factor due to gust at V_C.

An example of the use of this empirical formula in determining V_{Bmin} at high altitudes is shown in Fig. 15.5. The design V_B speed, shown in this figure, was selected to be higher than the minimum required to be consistent with the recommended climbout speeds for the airplane. The rationale was that this would minimize the slowdown distance to attain a desired turbulent air penetration speed V_{RA} during the climbout where turbulence may be significant due to cloud penetration.

15.5 Turbulent Air Penetration Speeds V_{RA}

Turbulent air penetration speeds may be defined as the maximum airspeed for rough air operation. In essence, it is an operational speed, shown in the AFM, that becomes the airspeed to which the pilot will slow down the airplane when very rough air is encountered or anticipated.

JAR ACJ 25X1517 defines V_{RA} as a speed that "(a) lies within the range of speeds for which the requirements associated with the design speed for maximum gust intensity, V_B, specified in JAR 25.335(d), are met, and (b) is sufficiently less than V_{MO} to ensure that likely speed variation during rough air encounters will not cause the overspeed warning to operate too frequently" (see JAR ACJ 25X1517).

JAR ACJ 25X1517 states, "V_{RA} should be less than V_{MO}—35 ktas."

As of 1994 there is no similar requirement in FAR 25, although the selected V_{RA} speeds for the AFM are approved by the FAA. In general, the V_{RA} speeds selected are as required by JAR 25.

As previously noted, V_{RA} speeds may be selected to be consistent with other operational speeds, such as climbout speeds. At high altitudes where Mach number becomes a consideration, the recommended speed in rough air must be selected

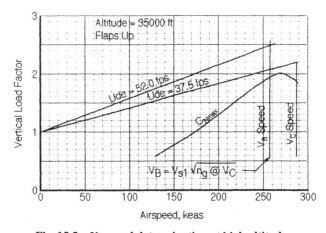

Fig. 15.5 V_B speed determination at high altitude.

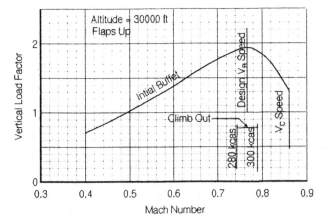

Fig. 15.6   Selection of V_{RA} speed at high altitude.

high enough to provide the margin for inadvertent gust-induced stall and low enough to protect the airplane structure against excessive overload. This speed is usually selected to be the Mach number M_{RA} at which the initial buffet curve peaks as noted in Fig. 15.6.

In practice the requirement for a minimum margin of 35 ktas is applied only to the altitudes where the speeds in rough air are stated in knots and not at altitudes where speeds are defined in terms of Mach number.

A list of the turbulent air penetration speeds are shown in Table 15.6 for some jet transports. In general, the selection of a single speed and Mach number for the recommended turbulent air penetration speed is advantageous to pilots due to the ease in remembering these speeds.[8]

Table 15.6   Turbulent air penetration speeds for U.S. jet transports[a]

Airplane	V_{RA}, kias	M_{RA}
Boeing 707	280	0.80
Boeing 727	280	0.80
Boeing 737	280	0.70
Boeing 747-200	280–290	0.82–0.85
Boeing 747-400	290–310	0.82–0.85
Boeing 757	290	0.78
Boeing 767	290	0.78
Convair 880	280	0.80–0.84
Douglas DC-8	280	0.80
Douglas DC-9-10	280	0.78
Douglas DC-9-30	285	0.79
Douglas DC-10	280–290	0.80–0.85
Lockheed L-1011	255–300	0.80–0.84

[a]The speeds listed were extracted from manufacturer or airline flight manuals.

Table 15.7   Recommended procedure for flight in severe turbulence

AIRPLANE FLIGHT MANUAL
NORMAL PROCEDURES

SEVERE TURBULENT AIR PENETRATION

Flight through severe turbulence should be avoided, if possible.
The recommended procedures for inadvertent flight in severe turbulence are:
1. Air Speed
   Approximately 290 KIAS or approximately 0.78 M, whichever is
   lower. Severe turbulence will cause large and often rapaid
   variations in indicated air speed. DO NOT CHASE THE AIR SPEED.
2. Yaw Damper - ENGAGED
3. Autopilot - DISENGAGE
4. Attitude
   Maintain wings level and the desired pitch attitude. Use
   attitude indicator as the primary instrument. In extreme
   drafts, large attitude changes may occur. DO NOT USE SUDDEN
   LARGE CONTROL INPUTS.
5. Stabilizer
   Maintain control of the airplane with the elevators. After
   establishing the trim setting for penetration speed, DO NOT
   CHANGE STABILIZER TRIM.
6. Altitude
   Allow altitude to vary. Large altitude variations are
   possible in severe turbulence. Sacrifice altitude in order to
   maintain the desired attitude and air speed. DO NOT CHASE
   ALTITUDE.
7. Thrust
   Engine ignition should be on. Make an initial thrust setting
   for the target airspeed. CHANGE THRUST ONLY IN CASE OF
   EXTREME AIR SPEED VARIATION.

15.5.1 Operational Experience

During the first 10 years of commercial jet transport operation, a significant
number of events involving extreme turbulence occurred.[8,9] These events raised
the question of V_{RA} speeds and how those aircraft are flown through turbulence.

The description of one of those encounters in Ref. 8 typifies many of the
occurrences during the pre-1964 period:

> 1) Pitch attitudes beyond 50-deg nose-up requiring both pilots holding full
> down elevator, 2) extreme up and down drafts during which pitch attitudes
> were steep enough to cause the horizon bar to disappear, 3) penetration of
> both low- and high-speed buffet boundaries, 4) very low minimum speeds,
> 5) maximum speeds beyond V_{MO}, possibly V_D, 6) reported inability to move
> either stabilizer or elevator in the high-speed dive despite the efforts of both
> pilots, 7) altitude excursion from 39,000 to 12,000 ft, and 8) peak load
> factors of $+3.2g$ and $-1.4g$.

These occurrences along with others led to the need to provide in the AFM a
procedure for flight in turbulence. These occurrences showed that the airplanes

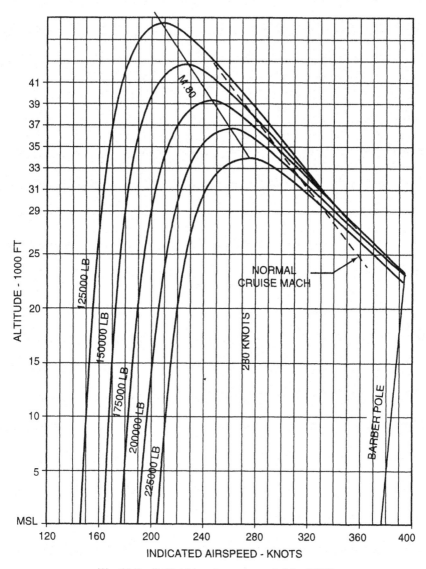

Fig. 15.7   Buffet boundary at $n_z = 1.5$ for 720B.

had a greater possibility to get into serious trouble as the result of inadvertent stall due to turbulence than to experience structural failure due to turbulence. An example of this procedure is shown in Table 15.7.

The selection of the recommended speed in rough air V_{RA} at higher altitudes where airspeed is monitored in terms of Mach number was given special attention. This may be seen in the examples shown in Figs. 15.7 and 15.8. The recommended Mach number in rough air M_{RA} is selected to maximize the margin for buffet, i.e., to provide the maximum gust encounter before entering the region where buffet occurs. (The definition of "barber pole" noted in Figs. 15.7 and 15.8 is the flight crew's name for the V_{MO}/M_{MO} limits shown on the cockpit airspeed indicators.)

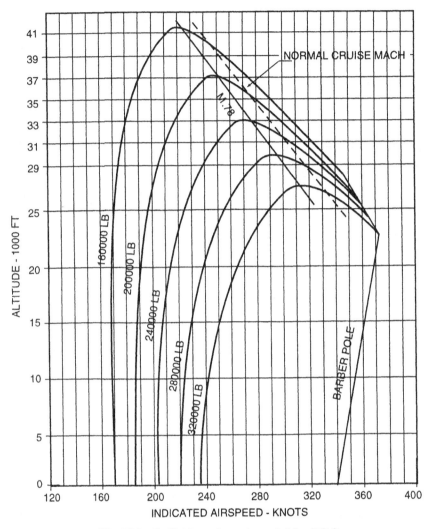

Fig. 15.8   Buffet boundary at $n_z = 1.5$ for DC-8.

15.6   Landing Gear Placards

Landing gear placard speeds are defined in two categories: 1) operating speeds: the maximum airspeeds at which the landing gears may be extended or retracted, which is normally designated as V_{LO}, and 2) maximum gear down speeds, which is normally designated as V_{LE}.

In general, the landing gear placards are selected from operational considerations but in all cases must comply with the following: "FAR/JAR 25.729(a)(2): Unless there are other means to decelerate the airplane in flight at this speed, the landing gear, the retracting mechanism, and the airplane structure (including wheel well doors) must be designed to withstand the flight loads occurring with the landing gear in the extended position at any speed up to 0.67 V_C."

Examples of these speeds are shown in Table 15.8.

Table 15.8   Landing gear extension and retraction placard speeds for two aircraft

	Airplane 1[a]	Airplane 2[b]
Minimum requirement per FAR/JAR 25.729(a)(2)		
Design V_C, gear up	350 keas	350 kcas
0.67 V_C	234.5 keas	234.5 kcas
Normal operation		
Gear extension	270 keas	270 kcas
Gear retraction	235 keas	270 kcas
Gear down/locked	320 keas	320 kcas/M 0.82
Dispatch landing gear down		
V_C/M_C	——	270 kcas/M 0.73
V_D/M_D	——	320 kcas/M 0.82

[a]Design speeds based on equivalent airspeeds.
[b]Design speeds based on calibrated airspeeds.

15.6.1   Dispatch—Landing Gear Down

To provide certification of commercial aircraft for revenue flights with gears locked down, the following consideration must be given in establishing operational placards.

A gear-down V_C speed must be established for dispatch operation. From this airspeed an upset margin is provided in determining a gear-down dispatch V_D speed. The airplane in example 2 shown in Table 15.8 required the speeds as noted in this table for dispatch operation, gear down.

An upset analysis using Eq. (15.1) was used to verify the dive speed shown in Table 15.8.

The dive speed chosen was the maximum airspeed for which the aircraft was designed with the gear down and locked for normal operation. The V_C speeds shown become the maximum airspeed for operating the airplane, gear down and locked in revenue service.

The airplane drag coefficients used in the solution to Eq. (15.1) were for a clean airplane plus the increment due to landing gear locked in the down position. The upset angle was the same as defined by Eq. (15.1), using the 7.5-deg required angle from the stabilized angle for flight at the airspeed from which the upset is initiated. A 20-s dive is assumed per FAR 25.335(b)(1).

15.7   Bird Strike Airspeed Considerations

Bird strike requirements are addressed in two separate sections of FAR. 25; namely 25.775, which establishes the requirements for windshields and windows, and 25.631, which applies to the empennage structure. The weight of the bird has changed over the years, but the airspeed requirements have been consistent from the prejet days to the modern commercial aircraft of today.

Consideration needs to be given to the design speed requirements for bird strike penetration. In general, FAR 25.775(b) is very specific on this point; i.e., "the velocity of the airplane (relative to the bird along the airplane's flight path) is equal to V_C at sea level, selected under FAR 25.335(a)."

JAR 25.631 states the following airspeeds need to be considered for design: "the velocity of the aeroplane (relative to the bird along the aeroplane's flight path) is equal to V_C at sea level or at 8000-ft, whichever is the more critical."

To put the preceding two criteria into perspective, consideration must be given to the basic problem of bird strikes, namely, the kinetic energy at the time of impact:

$$KE = 0.50(W_b/g)V^2 \qquad (15.17)$$

where KE is the kinetic energy (ft-lb); W_b is the weight of the bird (lb); g is the acceleration of gravity, 32.2 ft/s^2; and V is the true velocity of impact (ft/s).

True velocity is used in Eq. (15.17), and hence the more critical condition may be at the higher altitude depending on the selection of V_C speeds. This is shown in Table 15.9 for several selected V_C speeds. The purpose of the 8000-ft requirement in JAR 25 is to prevent the selection of a significantly low value of V_C at sea level to reduce the design requirement for bird impact, then selecting higher V_C speeds at altitude for operational purposes.

This is shown in the third set of airspeeds for the assumed airplane. The design V_C speeds are assumed to vary linearly from sea level to 10,000 ft. The low value at sea level was intentionally selected to show the rationale of the JAR 25.631 criteria.

As noted in Table 15.9, the true airspeeds are significantly higher as altitude increases; hence the question can be asked about bird strikes at higher altitudes. Again we must remember that bird strikes at the weight and airspeeds used in the

Table 15.9   Bird strike airspeeds

Altitude, ft	V_C speeds kcas	keas	ktas	Bird strike airspeeds, ktas
V_C defined as constant calibrated airspeed				
0	350	350.0	350.0	350
5000	350	347.8	374.6	——
8000	350	346.2	390.5	332[a]
V_C defined as constant equivalent airspeed				
0	——	350	350.0	350
5000	——	350	377.0	——
8000	——	350	394.8	336[a]
Assumed airplane with variable V_C[b]				
0	——	300	300.0	300
5000	——	325	350.1	——
8000	——	340	383.4	326[a]

[a]$0.85(V_C)$ ktas.
[b]The assumed airplane airspeeds demonstrate the rationale of the JAR 25.631 criteria.

criteria have been selected on a problematic basis and have been shown to provide adequate design for commercial airplane operation.

15.8 Stall Speeds

Stall speeds are the basis for determining various structural design speeds such as flap placard speeds, maneuver speeds V_A, and gust design speeds. The higher the stall speed for a given configuration, the higher the resulting structural design speed.

Stall speeds can be determined analytically using estimates of the stall lift coefficients for various flight configurations, i.e., flaps up or down, gear up or down, power on, etc. Initially these coefficients are usually provided by the aerodynamics department engineers. During flight test, stall speeds are verified, and updated stall lift coefficients are computed.

15.8.1 Historical Background for Stall Speeds

Before 1985 stall speeds were determined using a procedure that resulted in speeds obtained from flight tests at less than a 1-g load factor. These speeds, called FAR stall speeds, were used in the calculation of the various design speed requirements of FAR 25.

Stall lift coefficients are then calculated from these speeds as follows:

$$C_{L\,\text{stall}} = 295W / \left(S_w V_s^2\right) \qquad (15.18)$$

where W is the airplane gross weight (lb), S_w is the wing reference area (ft^2), and V_s is the FAR stall speed (keas).

15.8.2 Structural Design Philosophy for Stall Speeds

Structural design airspeeds have traditionally been determined per the requirements of FAR 25 (or CAR 4b), which had as its basis the so-called FAR stall speed that occurred at less than 1 g as discussed in Sec. 15.8.1.

In 1985 a change in philosophy on how flight test stall speeds should be demonstrated was enacted by the FAA; the traditional stall speeds will no longer be demonstrated, but only 1-g stalls will be verified in flight test. The purpose in the change in stall speed philosophy was essentially twofold.

1) Current aircraft were subjected to a very heavy buffet environment to minimize stall speeds.

2) The repeatability of flight test data had a lot to do with the skill of the flight test pilots. Very experienced pilots could repeat the results, but less skilled flight test pilots could not.

The change in philosophy in itself was not of major concern, but rather how the new 1-g stall speeds were to be used.

The approach by some is to be conservative and use the 1-g stall speeds in future calculations of design speeds per the requirements of FAR 25. This would in effect cause some loads such as trailing-edge flap loads to increase by as much as 12%, thus causing unnecessary added structural weight to the airplane. Before 1985, over 4000 airplanes had been designed and are currently in service using the older design philosophy, which has produced adequate structural integrity.

15.8.3 Definition of Stall Speed

Per FAR 25.103 the stall speed is defined as follows:

(a) V_s is the calibrated stalling speed, or the minimumsteady flight speed in knots, at which the airplane is controllable, with
   (1) Zero thrust at the stalling speed, or if the resultant thrust has no appreciable effect on the stalling speed, with engines idling and throttles closed;
   (2) Propeller pitch. . ..
   (3) The weight used. . ..
   (4) The most unfavorable center of gravity allowable.
(b) The stalling speed V_s is the minimum speed obtained as follows
   (1) Trim the airplane for straight flight at any speed not less than $1.2V_s$ or more than $1.4V_s$. At a speed sufficiently above the stall speed to ensure steady conditions, apply the elevator control at a rate so that the airplane speed reduction does not exceed one knot per second.
   (2) Meet the flight characteristics provisions of FAR 25.203.

Per the requirements of FAR 25.201, stall demonstrations shall be accomplished as follows:

(a) Stalls must be shown in straight flight and in 30 degree banked turns with
   (1) Power off; and. . ..
   (2) The power necessary to maintain level flight at 1.6 V_{s1} (where V_{s1} corresponds to the stalling speed with flaps in the approach position, the landing gear retracted, and maximum landing weight). . ..
(d) Occurrence of stall is defined as follows:
   (1) The airplane may be considered stalled when, at an angle of attack measurably greater than that for maximum lift, the inherent flight characteristics give a clear and distinctive indication to the pilot that the airplane is stalled. Typical indications of a stall, occurring either individually or in combination, are:
      (i) A nose down pitch that cannot be readily arrested;
      (ii) A roll that cannot be readily arrested; or
      (iii) If clear enough, a loss of control effectiveness, and abrupt change in control force or motion, or a distinctive shaking of the pilot's controls.
   (2) For any configuration in which the airplane demonstrates an un-mistakable inherent aerodynamic warning of a magnitude and severity that is a strong and effective deterrent to further speed reduction, the airplane may be considered stalled when it reaches the speed at which the effective deterrent is clearly manifested.

15.8.4 Stall Speeds as Defined in JAR 25.103

(b) The stall speed V_s is the greater of
   (1) The minimum calibrated airspeed obtained when the aeroplane is stalled (or the minimum steady flight speed at which the aeroplane is controllable with the longitudinal control on its stop) as determined when the manoeuvre prescribed in JAR 25.201 and 25.203 is carried out with an entry rate not exceeding 1 knot per second; and
   (2) A calibrated airspeed equal to 94% of the one-g stall speed, V_{s1g} determined in the same conditions.
(c) The $1g$ stall speed, V_{s1g}, is the minimum calibrated airspeed at which the aeroplane can develop a lift force (normal to the flight path) equal to its

weight, whilst at an angle of attack not greater than that at which the stall is identified.

15.8.5 Implication of Stall Speed Selection

The ratio of FAR stall speeds to 1-g stall speeds varied from 0.91 to 0.97 on many of the aircraft on which the author has worked.

Depending on the flap configuration, the design loads would increase from 7 to 17% if placard speeds defined by Eqs. (15.8) and (15.9) were based on V_{s1g}. These loads would be higher than the maximum landing flap position that may be protected by a flap retractor and would become critical for design of the trailing-edge flaps and backup structure.

15.8.6 Concluding Remarks on Stall Speeds

The importance of stall speeds to the structural loads engineer cannot be emphasized enough. Because of the compounding effect on design speeds, overly conservative stall speeds must be avoided. Higher operating empty weight resulting from conservative airspeeds will impact the airplane performance parameters by reduced payload, range, or field length capability.

References

[1]Anon., "Airplane Airworthiness, Transport Category," Civil Air Regulations, Part 4b, amended Dec. 31, 1953.

[2]Anon., "High Speed Characteristics," FAA Advisory Circular 25.253-1, Nov. 1965.

[3]Coleman, T. L., Copp, M. R., and Walker, W. G., "Airspeed Operating Practices of Turbine-Powered Commercial Transport Airplanes," NASA TN D-744, April 1961.

[4]NASA Staff, "Airspeed Operating Practices of Turbine-Powered Commercial Transport Airplanes," NASA TN D-1392, Oct. 1962.

[5]Anon., "Airplane Airworthiness, Transport Category Airplanes," Pt. 25, Federal Aviation Administration, revised Jan. 1, 1994.

[6]Pearson, H. A., "Study of Inadvertent Speed Increases in Transport Operation," NACA TN 2638, March 1952.

[7]Bisplinghoff, R. L., Ashley, H., and Halfman, R. L., Aeroelasticity, Addison–Wesley, Reading, MA, 1955.

[8]Soderlin, P. A., "Flight Standards Bulletin No. 8-63," Northwest Airlines, Nov. 1963.

[9]Stapleford, R. L., and DiMarco, R. J., "A Study of the Effects of Aircraft Dynamic Characteristics on Structural Loads Criteria," U.S. Dept. of Transportation, Nov. 1978.

16
Airspeeds for Structural Engineers

This chapter was prepared to address the problem of understanding airspeeds by engineers without an aeronautical background. We hear the terms calibrated airspeed, equivalent airspeed, indicated airspeed, and finally true airspeed. What do these terms mean and how are they used?

16.1 Relationship of Lift to Airspeed

First, let us consider the airplane in simple terms in which lift is defined in terms of a nondimensional coefficient, dynamic pressure, which is a function of airspeed and wing reference area:

$$\text{Lift} = C_L q S_w \qquad \text{(lb)} \qquad\qquad (16.1)$$

where C_L is the lift coefficient, S_w is the wing reference area (ft^2), and q is the dynamic pressure (lb/ft^2).

16.2 Equivalent Airspeed

Dynamic pressure can be defined in several ways. Consider the definition in terms of true airspeed as shown in Eq. (16.2):

$$q = \tfrac{1}{2}\rho V_T^2 \qquad \text{(lb/ft}^2\text{)} \qquad\qquad (16.2)$$

where ρ is the density of air (slug/ft^3), and V_T is the true airspeed (ft/s).

Rewriting Eq. (16.2) in terms of sea level values, the dynamic pressure at sea level is defined by Eq. (16.3):

$$q = \tfrac{1}{2}\rho_0 V_0^2 \qquad \text{(lb/ft}^2\text{)} \qquad\qquad (16.3)$$

where ρ_0 is the density of air at sea level (slug/ft^3), and V_0 is the true airspeed at sea level (ft/s).

Using the value of the density of air at sea level and the conversion of the airspeed units from feet per second to knots, the dynamic pressure at sea level as derived from Eq. (16.3) becomes

$$q = V_e^2 / 295 \qquad \text{(lb/ft}^2\text{)} \qquad\qquad (16.4)$$

where V_e is defined as knots equivalent airspeed. The conversion units used in Eq. (16.4) are $\rho_0 = 0.0023769$ slug/ft^3,[1] feet per second = (88/60)(6080.27/5280) = 1.689 kn,[2] and knots = nautical miles per hour.

Equivalent airspeed is, in essence, an airspeed at sea level that would result in the same dynamic pressure q experienced by the airplane flying at altitude at its true airspeed. A constant equivalent airspeed will give the same lift at all altitudes for the same gross weight and load factor.

16.3 Relationship Between Equivalent Airspeed and True Airspeed

Equating Eqs. (16.2) and (16.3),

$$V_0 = V_T(\rho/\rho_0)^{\frac{1}{2}} \tag{16.5}$$

which becomes, in terms of equivalent airspeed

$$V_e = V_T(\rho/\rho_0)^{\frac{1}{2}} \tag{16.6}$$

Because of the reduction in density of air with altitude, true airspeed must increase to provide constant lift. This may be seen by combining Eqs. (16.1) and (16.2):

$$\text{Lift} = \left(\tfrac{1}{2}\rho V_T^2\right)C_L S_w \tag{16.7}$$

16.4 Indicated Airspeed

The speed that the pilot reads on the airspeed indicator in the cockpit is called indicated airspeed. It differs from calibrated airspeed only in terms of instrument and static source error, which may be a function of airplane flight attitude, Mach number, and flap position.

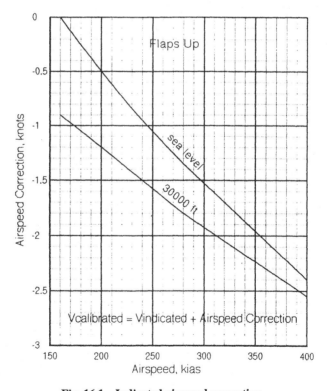

Fig. 16.1 Indicated airspeed correction.

An example of the correction of indicated airspeed to calibrated airspeed is shown in Fig. 16.1 for a typical commercial jet installation. Only two altitudes are shown in this example.

16.5   Calibrated Airspeed

So how does calibrated airspeed relate to equivalent or true airspeed as shown in Eqs. (16.2–16.4)?

The relationship between these airspeeds is a function of the pressure ratio and Mach number at altitude, as shown in Eqs. (16.8–16.10).

Given Mach number and altitude, the calibrated airspeed may be calculated from Eq. (16.8):

$$V_C = a_0\sqrt{5(\{\delta[(1 + 0.2M^2)^{3.5} - 1] + 1\}^{1/3.5} - 1)} \qquad (16.8)$$

Given calibrated airspeed and altitude, the Mach number may be calculated from Eq. (16.9):

$$M = \sqrt{5(1/\delta\{[1 + 0.2(V_C/a_0)^2]^{3.5} - 1\}^{1/3.5} - 5)} \qquad (16.9)$$

Given calibrated airspeed and Mach number, the pressure ratio may be calculated from Eq. (16.10), and hence the altitude for the condition may be determined:

$$\delta = \frac{[1 + 0.2(V_C/a_0)^2]^{3.5} - 1}{[1 + 0.2M^2]^{3.5} - 1} \qquad (16.10)$$

where a_0 is the speed of sound at sea level, 661.287 kn for a standard day; M is the Mach number at altitude; V_C is the calibrated airspeed (kcas); and δ is the pressure ratio at altitude.

Table 16.1   Standard atmospheric parameters obtained from Ref. 1

For H < 36,089 ft	For H > 36,089 ft
$H = k_1(1 - \delta^{0.19026})$, ft	$H = 36,089 + k_3 \ln(0.22336/\delta)$, ft
$\delta = (1 - H/k_1)^{5.2561}$	$\delta = 0.22336e^{-x}$
	where $x = (H - 36089.24)/k_3$
$\sigma = (1 - H/k_1)^{4.2561}$	$\sigma = \delta/(0.75187)$
$a_e = 661.287\delta^{0.5}$, keas	$a_e = 661.287\delta^{0.5}$, keas

where
  $k_1 = 518.688/0.0035662 = 145445.57$
  $k_3 = (1716.5)(389.988)/(32.17405) = 20806.03$
  $H =$ altitude, ft
  $\delta =$ pressure ratio
  $\sigma =$ density ratio
  $a_e =$ speed of sound, keas

Table 16.2   Calibrated and true airspeed vs equivalent airspeed

| Altitude, ft | Airspeed[a] | | | |
	kcas	keas	ktas	Mach no.
Sea level	250	250.0	250.0	0.378
10,000	250	248.1	288.7	0.452
20,000	250	245.2	335.9	0.547
25,000	250	243.3	363.4	0.604
30,000	250	240.8	393.7	0.668
35,000	250	237.8	427.2	0.741
40,000	250	234.2	472.0	0.823
Sea level	350	350.0	350.0	0.529
10,000	350	345.1	401.5	0.629
20,000	350	337.9	462.9	0.754
25,000	350	333.2	497.7	0.827
30,000	350	327.6	535.5	0.909
35,000	350	320.8	576.4	1.0

[a]Where kcas = knots calibrated airspeed, keas = knots equivalent airspeed, and ktas = knots true airspeed (actual speed over the ground in still air).

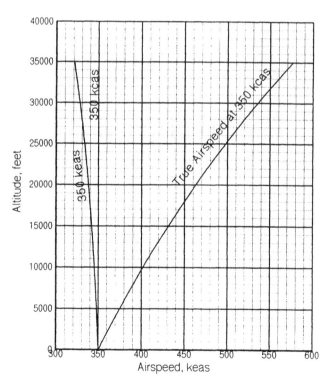

Fig. 16.2   Calibrated and true airspeeds vs equivalent airspeeds.

Knowing Mach number and altitude, one can calculate equivalent airspeeds from Eq. (16.11):

$$V_e = a_e M \tag{16.11}$$

where a_e is the speed of sound at altitude (keas) (see Table 16.1), and M is the Mach number.

16.6   True Airspeed

True airspeed can be calculated from equivalent airspeed using Eq. (16.12), which is obtained from Eq. (16.6):

$$V_T = V_e \sigma^{-\frac{1}{2}} \tag{16.12}$$

where σ is the density ratio at altitude, ρ/ρ_0 (see Table 16.1).

16.7   Variation of Equivalent Airspeed and True Airspeed with Altitude

For a constant calibrated airspeed with altitude, the equivalent airspeed will decrease with increasing altitude, whereas the true airspeed will increase with increasing altitude as shown in Table 16.2.

The variations of constant calibrated and equivalent airspeeds with altitude are shown in Fig. 16.2. The true airspeed as shown is for a constant 350-keas airspeed.

References

[1] Anon., "Manual of the ICAO Standard Atmosphere, Calculations by the NACA," NACA TN 3182, May 1954.

[2] Eshbach, O. W., *Handbook of Engineering Fundamentals,* 1st ed., Wiley, New York, 1945.

Index

CPSIA information can be obtained
at www.ICGtesting.com
Printed in the USA
BVOW04*1644290117

474289BV00014B/8/P

9 781563 471148